REASONING SKILLS FOR HANDLING CONFLICT

David W. Felder, Ph.D.

Wellington Press

ISBN 9781575017754

Wellington Press
9601-30 Miccosukee Road
Tallahassee, FL 32309-9662
www.booksuprint.com
E-Mail: booksuprint@comcast.net

Reasoning Skills for Handling Conflict
David W. Felder, Ph.D.

Contents

Introduction

The materials in this book are used in situations where there is disagreement. One situation is when you disagree with someone else, another is when you witness other people having a disagreement and you want to decide what views to accept. Whichever situation you find yourself in, the problem is the same. There is disagreement and you want to end up with a situation of agreement. You want to go from a situation of conflict to one of consensus.

The first step in attaining the goal of agreement is to understand the nature of the disagreement. Each type of disagreement is handled a different way. Knowing the type of disagreement allows you to know what tools to use in coming to agreement. The first chapter of this book presents methods for determining the type of disagreement.

Other chapters of this book cover methods for handling each type of disagreement. Chapter two shows how to handle verbal disagreements with definitions. Chapter three presents methods of handling factual disagreements with arguments and chapter four how to avoid fallacies when presenting arguments. Chapter five focuses on methods of conflict resolution for handling conflicts of interest, and chapter six examines methods for handling conflicting moral claims. Chapters seven through ten present the methods of symbolic logic for handling factual disputes.

The skills covered in this book include the ability to communicate your viewpoints and the ability to understand other people's viewpoints. In cases where it is possible to reach agreement, you will learn how to go from conflict to consensus. In cases where agreement is not possible, you will be able to understand the nature of the disagreement. These are important skills for personal life, employment, and the survival of life on our planet. Skills in handling conflict are especially important today because when people lack these skills they turn too readily to violence. While presenting these skills which are important to life, the text also covers most of the reasoning skills that are tested on standardized achievement tests.

This text will increase your reasoning power. The ability to reason is a real power. You can use this power to convince people to do what you want them to do, or to free yourself from the control of others. Those who have this power are better off then those who do not. There are several specific benefits of having reasoning power. You will enjoy greater self-confidence knowing that you can hold your own in reasoning ability. Second you will be able to see through tricks that con artists pull on the unsuspecting. A third benefit is that you will be able to trick others. Reasoning power is a real power and like other powers it can be used for good or evil. You are being entrusted with this power and it is up to you to choose how to use it. May the power be with you.

Chapter 1:

DETERMINING THE TYPE OF DISAGREEMENT

Instructional Purpose

It is important to understand the types of disagreements people have because the type of disagreement determines what is needed in order to come to agreement. If a dispute is factual, then one might attempt to settle it by the process of argumentation; but if a dispute is one of attitude, nothing can be done unless one analyzes the factual background of the negative or positive attitude. If a dispute is one that is merely verbal, it can sometimes be solved by defining words. Conflicts of interest and moral disputes are each handled in different ways. Because different types of disputes are settled by different methods, a good first step is to identify the type of dispute.

Instructional Objective

Upon completion of this unit, the student will be able to classify disagreements as ones of attitude, fact, verbal disagreements, conflicts of interest or moral disputes.

Sample Test Item

Directions

Indicate what is true of each example using the following choices:

A)   The disagreement is one of attitude.
B)   The disagreement is one of facts.
C)   The disagreement is verbal.
D)   The disagreement is a conflict of interest.
E)   The disagreement is a moral dispute.

Example

Mr. X: "She is very devoted to the cause."
Mr. Y: "She is a fanatic for the cause."

Answer

A

Behavioral Objective

After completing this chapter, the student should be able to:

1.   Define and recognize examples of the following.
      Attitude Differences
            slanted favorably
            slanted unfavorably
            neutral

Functions of Language
    informative
    expressive
    directive
    ethical
    performative
    ceremonial
Types of Disagreement
    attitude
    factual
    verbal
    conflict of interest
    moral dispute

2.   State how people having each type of disagreement might come to agreement.

Organizer - Spotting Attitudes

Before one can spot whether there is disagreement in attitude, one must first know how to spot attitudes.

CONCEPTS: Slanted Favorable, Slanted Unfavorable, Neutral

- Slanted favorable = If a sentence is slanted favorably, then the reader can tell that the author likes the sentence's subject.

- Slanted unfavorable = If a sentence is slanted unfavorably, then the reader can tell that the author dislikes the subject of the sentence.

- Neutral = If a sentence is neutral then the reader cannot tell whether the author likes or dislikes the subject of the sentence.

Organizer - Denotation and Connotation

A person can often spot attitudes by being sensitive to the connotative meaning of words. So first we will discuss the two types of meanings words have: denotation and connotation.

CONCEPTS: Denotation and Connotation

Examine the set of sentences below and observe how the statements in the set differ.

2

Denotation & Connotation
I am a peace officer.
You are a policeman.
He is a flatfoot.

The terms "*peace officer*," "*policeman*," and "*flatfoot*" all denote or refer to the same people; that is, people who work for a police department. Thus, these terms have the same denotative meaning because they refer to the same subjects.

The terms "*peace officer*," "*policeman*," and "*flatfoot*" differ in their emotive meaning. Each term shows a different feeling toward the person referred to. The term "peace officer" has a positive connotation and the term "flatfoot" has a negative connotation. The term "policeman" is neutral and has no strong connotation, either positive or negative.

The irregular conjugation "I am, you are, he is, etc." allows us to notice immediately whether words are positive or negative. When describing oneself, positive words are usually used. In describing a person to his face, neutral words or better are used. When a person is not present, negative words are more easily used. In determining whether a word is positive, neutral, or negative, you might ask whether a person would use the word to describe himself, someone else to her face, or would use the word only when the person is not present.

Sample Test Item

Directions

Indicate what is true of this example using the following choices:

A) Approval
B) Disapproval
C) Neither

Example

"The tramp sneaked behind the barn."

Answer

B

3

Procedure	1.	Examine the words used. A writer has options regarding what words are used. Some words have positive connotations and other words have negative connotations. Some words are complimentary and other words are derogatory. (In this case, the words "tramp" and "sneaked" have derogatory connotations).
	2.	Examine what is mentioned and what is not mentioned. A writer has options on what he or she picks to mention.
	3.	Examine what is suggested in addition to what is said. A person might make suggestions which go beyond the facts. For example, "in a strictly legal sense he did nothing wrong" implies that he did do something wrong. This involves reading between the lines.

Basically, a person shows his or her attitudes in the choices he or she makes. A writer has options on what words to use and what to mention. Consider all the options the writer had and ask whether he chose favorable, unfavorable, or neutral options. One might imagine a list of words with the same denotations which might be used by an author with positive words at the top of the list and the most negative words at the bottom. With this model in mind, ask yourself whether the author chose a word which would fall at the top, bottom or middle of the list.

Spotting Slanted Coverage

To spot slanted coverage examine:

1. ... the words and their connotations
2. ... what is said and what is not said
3. ... what is suggested vs. what is said

CONCEPTS: Favorable or Unfavorable Slant, Neutral Coverage

If it is possible to tell that an author approves of the subject he is writing about without his stating directly that he approves, then the writing is slanted favorably toward that subject.

If it is possible to tell that an author disapproves of the subject he is writing about without his stating directly that he disapproves, then the writing is slanted unfavorably.

If it is not possible to tell whether a writer approves or disapproves of the subject about which he is writing, then the coverage is neutral.

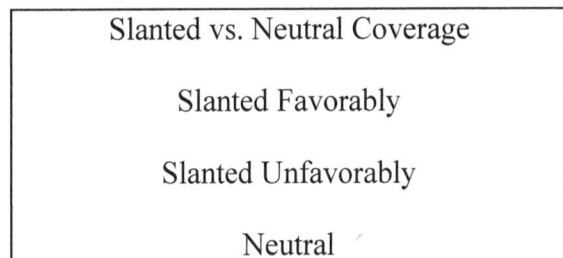

Slanted vs. Neutral Coverage
Slanted Favorably
Slanted Unfavorably
Neutral

Directions: Indicate what is true of each example using these choices:
- A) Slanted favorably toward the subject of the sentence.
- B) Slanted unfavorably toward the subject of the sentence.
- C) Neutral toward the subject of the sentence.

1. He was rash to jump into the water after the child ....____

2. He was brave to jump into the water to save the child. ....____

3. He jumped into the water and swam toward the child. ....____

4. He attends many meetings for the movement...____

5. He is a fanatic for the cause. ....____

6. He is very devoted to improving conditions. ....____

7. She is very lively at parties. ....____

8. She moves around a lot at parties. ....____

9. She goes wild at parties. ....____

10. He has a lot of nerve crashing parties. ....____

11. He has a lot of courage going to unfamiliar parties. ....____

12. He goes to parties he is not invited to. ....____

13. She is even heavier than he is. ....____

14. The suspect squandered the funds. ....____

15. Wife to friend: "My husband was sober last night." ....____

16. The driver, pushing the legal speed limit, despite the rain, hit him. ....____

17. The mechanic had oil on his hands. ....____

18. The union is trying to force through a pension plan. ....____

19. Although criticized from many quarters, she stood her ground. ....____

20. The American flag was conspicuously absent from the speakers' platform. .... ____

Exercise 1 - 2

Directions: Examine newspaper or magazine articles for examples of slanted news coverage. Ask yourself whether you can tell the attitude of the author of an article. Bring in an example in which you can tell the attitude of an author and explain what in the article shows the author's attitude.

1. Words used _____

_____

_____

2. What is and is not mentioned _____

_____

_____

_____

3. What is suggested vs. what is said _____

_____

_____

_____

_____

4. Your overall assessment on slant _____

_____

_____

_____

Organizer - Three Main Functions of Language

> Before we get to types of disagreement, it is desirable to learn to spot functions of language. Factual disagreements are usually expressed in informative language, attitudinal disagreements in expressive language, and so on.

CONCEPT: Informative Language

Definition	Informative language includes utterances which are true or false.
Example	"The planet Jupiter is larger than the Earth."
Procedure	To test for Informative, ask "Is the utterance true or false?"

CONCEPT: Expressive Language

Definition	Expressive language includes utterances in which a person expresses his or her <u>approval</u> or <u>disapproval</u> of something.
Example	"I hate the smell of insect spray."
Procedure	To test for Expressive, ask "Can you tell whether the speaker approves or disapproves of something?"

CONCEPT: Directive Language

Definition	Directive language includes utterances which tell a person to do something or not to do something.
Example	"Close the door."
Procedure	To test for Directive, ask "Does the utterance tell someone to do something or not to do something?"

Note:
To remember these functions of language, you might associate a few words with each one.

The Main Function of Language
Informative   (true/false)
Expressive (approval/disapproval)
Directive   (do it/don't do it)

Note:
Remember the following to avoid confusing informative and expressive. One difference between informative and expressive is that when one makes an informative utterance, he or she is telling others some fact about the world, whereas when one makes an expressive utterance one is telling others about oneself, e.g., what is liked or not liked. One test to perform is to ask yourself whether you have learned about the speaker or about a reality outside the speaker.

Note:
It is acceptable to have several functions of language used in the same discourse, providing that a person realizes he or she is switching form one to the other. When one makes a factual claim, one needs to support the information with other solid information rather than emotion.

Name _____

Exercise 1 - 3

Directions: Examine the paper below and list next to each numbered sentence the function or functions of language which are present. Then evaluate the paper. The author of the paper was told to list a belief she disagreed with and then to try to convince others that they should also reject that belief.

Green People Should Be Treated As Equals

(1) I made this statement many years ago. (2) The reason for this foolish idea was the fact that I was too young to really know what they were doing to themselves and to the society in which they lived. (3) I used to laugh at the clothes they wore and how they walked and talked. (4) This was really funny to me because everyone else laughed which made it even funnier.

(5) People think that this country has a drug problem. (6) They need to look again. (7) This country has a problem all right, a Green epidemic.

(8) The number of Green People is growing more and more each day. (9) If the world doesn't wake up in time there won't be a decent man or woman alive by the year 2,000. (10) Soon things will get so out of hand until people will be choosing their own color.

(11) It is not too late. (12) People should stop laughing and start some serious thinking about this matter. (13) If they were treated like dogged infested disease maybe the number of Green People would be reduced.

(14) Now our society is slowly changing form. (15) Instead of the normal colors, we see Green People everywhere. (16) When will this nightmare end? (17) It's got to be stopped and we are the only ones that can do it. (18) Now that I am a little older, I have come to realize how repulsive they really are. (19) Their activities aren't funny any more. (20) They turn my stomach.

It is possible to go from one function to another. A person might, for example, get a person to close a door by uttering a directive, "Close the door," or he might get a person to do the same act by giving an informative utterance, "It's cold in here." Consider the following interchange"

Bob:	Close the door
Joe:	I don't want to.
Bob:	It's getting cold in here. I have a cold and it's getting worse.
Joe:	No! It's not cold.
Bob:	The thermometer says 30 and the water in the pitcher is frozen. You have an icicle hanging from the tip of your nose.
Joe:	O.K., I'll close the door.

Exercise 1 - 4   Directions:

1. Think of information which might affect people's attitudes toward people of another ethnic group. Imagine a Mr. Guff who hates Bunts. How might Mr. Guff's attitude toward people of the Bunt ethnic group be changed?

2. Imagine that you are the leader of a nation and that you want your country to invade another nation. Write a speech that is designed to make people in your country hate people in the other country enough so they will want to kill them.

Organizer -   Some Other Functions of Language

There are other functions of language besides the informative, expressive, and directive functions. There are as many functions of language as there are things people do with language. Only a few of the additional functions of language are defined below.

CONCEPT: Performative Language

Utterances which perform actions, such as apologizing and promising, come under the performative function of language. A person is performing an action with words.

Example   "I promise."

Procedure   To test for Performative, ask yourself . . . Is the person making the utterance performing an action? If

11

she is, then you can ask whether she did it, i.e., did she apologize?

CONCEPT: Ethical Language

	Utterances which claim that a thing, person, action or situation ought or ought not to exist. Any claim that something is morally good or morally evil involves ethical language.
Example	"It is wrong to steal."
Procedure	To test for Ethical language, ask yourself . . . Is the person making the utterance claiming that something ought or ought not to exist? If so, then the utterance involves ethical language.

CONCEPT: Ceremonial Language

	Utterances which are part of ritual behavior. Greetings and salutations are part of the ceremonial function of language.
Example	"Hello."
Procedure	To test for Ceremonial, ask yourself, . Is the expression something you say simply because it is the custom? If so, then the expression is ceremonial.
Note:	To remember these last three functions of language, you might just remember an example of each.

Other Functions of Language
Performative Language - "I promise" Ethical Language - "Lying is wrong" Ceremonial Language - "Hello"

Note:	Remember the following to avoid confusing directive and performative. One difference between directive and performative is that when one uses directive language, he is getting someone else to do something; when one uses performative language, she is performing an action.

Exercise 1 - 5   **Name**_____

Directions: Indicate what s true of each example using these choices:

        A)   Informative Language
        B)   Expressive Language
        C)   Directive Language
        D)   Performative Language
        E)   Ethical Language
        F)   Ceremonial Language

1.   I like orange juice. ....____

2.   Do not waste your food. ....____

3.   The postman is late. ....____

4.   Hello. ....____

5.   Take a pill after each meal and before going to bed. ....____

6.   I apologize. ....____

7.   Your flight leaves in twenty minutes. ....____

8.   A person is never too old to learn. ....____

9.   Take your umbrella with you today. ....____

10. A child should always obey his or her parents.. ____

11. I promise to pay you back next week. ....____

12. George was here at three this afternoon. ....____

13. The President has made many promises of that sort before. ....____

14. This package is sold by weight, not volume. ....____

15. Fill out this questionnaire. ....____

16. It is wrong to steal. ....____

17. I'd love to go to a restaurant for dinner. ....____

18. Close the door. ....____

19. I don't like beets. ....____

20. Passing this bill is the right thing to do. ....____

Now that we know how to spot attitudes, we can go on to learning how to tell whether a disagreement is one of attitude, facts, a verbal dispute, an ethical disagreement, or a conflict of interest.

CONCEPT: Attitude Disagreement

Definition	An Attitude Disagreement is present when one person has an attitude toward a subject which is different from the attitude a second person holds.

Example

Sil: Suzy is lively at parties.
Sal: Suzy goes wild at parties.

Procedure

To test for Attitude Disagreement . . .

1. Spot the attitude in the first statement. It is a positive attitude.
2. Spot the attitude in the second statement. It is a negative attitude.
3. See if the attitude in the first statement is different from the attitude in the second statement. In this case, one is positive and the other is negative, so there is a disagreement in attitude.
4. Ask, "Does one utterance show one attitude toward the subject which is different from the attitude shown in the other utterance?"
5. Very often with an attitude difference, people are saying the same thing, but they are saying it differently.

CONCEPTS: Factual Disagreement

Definition	A factual disagreement is present when two or more people make assertions which cannot be true at the same time.

Example

Sil: Mr. Green is a college graduate.
Sal: Mr. Green never went to college.

Procedure

To test for Factual Disagreement, ask yourself . .

1. "Can both statements be true at the same time?" If the disagreement is factual, then the two statements cannot both be true.

2. Replace any words having strong connotations with neutral words. If one person says "He is a peace officer," and the other says, "He is a flatfoot," both statements become the statement, "He is a policeman," when the neutral term "policeman" is substituted for "peace officer" and "flatfoot."

CONCEPT: Verbal Disagreement

Definition

A Verbal Disagreement is present when two or more parties believe that they have a disagreement, yet the disagreement depends entirely on the words used.

Example

Sil: Ancient Athens was a democracy because people in Athens elected their leaders.

Sal: Ancient Athens was not a democracy because seven out of eight individuals were slaves in Athens.

Procedure

To test for Verbal Disagreement . . .
1. See if you can identify a word which could be changed or re-defined in a way that would end the dispute.
2. See if there are different meaning being used for a word by the following process:
   (a) Think of a definition of terms that would work for the first statement or speaker.
   (b) Think of a definition of terms that would work for the second statement or speaker.
   (c) See if there are two different definitions of terms being used by the speakers.

CONCEPT: Ethical Disputes

Definition

An Ethical Dispute is present when two or more parties disagree over ethical statements. Typically one person states that a particular thing is good, while another states that the same thing is bad; one believes that something ought to exist, another that it ought not to exist.

Example

Sil: Wealth ought to be divided equally.
Sal: Wealth ought not to be divided equally.

Procedure	To test for Ethical Disputes . . .
	1. See if people are using ethical language and making ethical claims.
	2. See if the people are assuming ethical statements which they are not stating directly. If so, state the assumed ethical claims.
	3. Note the ethical statements which each person is committed to, and see whether both could be accepted by one person. A person cannot accept that one thing is both good and evil at the same time. If both views cannot be accepted, then there is an Ethical Dispute.

CONCEPT: Conflict of Interest

Definition	A Conflict of Interest is present when two or more parties want something which cannot be either divided or shared, so that the attainment thereof by one party precludes attainment by the other. Conflicts of Interest are usually conflicts over either the possession or use of desired objects.
Example	Employer: I want to pay you workers only ten dollars an hour. Employee: We want at least fifteen dollars an hour.
Procedure	To test for Conflict of Interest . . .
	1. See if there is a statement which indicates what each person wants to either possess or use.
	2. See whether the attainment of what one person wants would preclude the attainment of what another person wants. If it would, then there is a conflict of interest.

16

Directions: Indicate what is true of each example using these choices:

A)   Attitude Disagreement
B)   Factual Disagreement
C)   Verbal Disagreement
D)   Ethical Disagreement
E)   Conflict of Interest

1.   "John is 5' 10" tall."
      "John is 6' 1" tall." ....____

2.   "Mary almost met her sales quota."
      "Mary failed to meet her sales quota." ....____

3.   "Passing this bill is the right thing to do."
      "Passing this bill is the wrong thing to do." ....____

4.   John has a new hair cut.
      John has the same old hair cut he always wears. ....____

5.   "We want seventeen dollars an hour for our labor."
      "As manager I can give you no more than fifteen dollars an hour." ....____

6.   "Mr. Jones is thrifty."
      "Mr. Jones is cheap." ....____

7.   "Harry finally got rid of that old Ford of his and bought a new car. He's
      driving a Chevy now."
      "No, Harry did not buy a new car. That Chevy is a good three years
       old." ....____

8.   "The sun revolves around the earth."
      "The earth revolves around the sun." ....____

9.   "I like this class."
      "I don't like this class." ....____

10. "That's my pen, because I saw it first.
      "No, it's mine. I picked it up." ....____

11. "The earth has been in existence for only a hundred million years."
      "The earth has been in existence for a hundred million years." ....____

12. "Karen lives a long way from campus. I walked out to see her the other
       day, and it took me nearly two hours to get there."

"Karen does not live a long way from campus. It took me only ten
minutes to drive to her apartment." ....____

13. "The sending of United States troops for UN Peacekeeping is wrong.
"It is right that the United States send troops for UN Peacekeeping.........................____

14. "Senator Gray is a genuine liberal. She votes for every progressive
measure that comes before the legislature."
"Senator Gray is not liberal. She hardly gives money to any causes."......................____

15. "Joe missed the ball."
"Joe almost hit the ball." ....____

16. "With an income of twenty thousand a year, the Smiths are poor."
"With an income of twenty thousand a year, the Smiths are not poor." ....____

17. The glass of wine is half full.
The glass of wine is half empty. ....____

18. "A tree falling in the wilderness with nobody around to hear will
produce no sound. There can be no auditory sensation unless
someone actually senses it."
"No, whether anyone is there to hear it or not, the crash of a falling
tree will set up vibrations in the air and will therefore produce a
sound." ....____

19. "That is my wallet you have in your hand."
"It's my wallet now." ....____

20. "Mercury is smaller than Venus in circumference.
"Mercury is larger than Venus in circumference." ....____

Chapter 2:

SETTLING VERBAL DISPUTES WITH DEFINITIONS

Instructional Purpose

Definitions aid in the settling of verbal disputes, while arguments aid in the settling of factual disputes. After determining the type of disagreement, you know whether the disagreement is one that can be settled by definitions. If it can be settled by definitions, then you must know how to formulate adequate definitions in order to settle the dispute.

Instructional Objective

Students will be able to demonstrate that a given definition is adequate.

Sample Test Item

Directions

Indicate whether the following definition is adequate or whether it is too broad, too narrow, inconsistent, or circular by writing your answer in the answer column.

Example

quart = fraction of a gallon

Answer

too broad

Behavioral Objective

After completing the chapter, the student should be able to:

1. Define and recognize examples of the following:
    Types of Definitions:
        Definitional Report
        Definitional Proposal
    Errors in Definitional Report
        Too Broad
        Too Narrow
        Too Broad and Too Narrow
        Inconsistent
        Circular

2. State the relationship between intension and extension.

ORGANIZER - **Report vs. Proposal**

When people have a merely verbal disagreement, they need to come to agreement on the meaning of the word which is causing their disagreement. One way to agree on the meaning of a word is for someone to propose a meaning; the other way is to report how a word is used. A definitional report can be correct or incorrect.

CONCEPT - Definitional Report vs. Definitional Proposal

A definitional report simply tells how a word is generally used. It should be neutral and report objectively and accurately the current or past usage of the word. A dictionary gives definitional reports.

A definitional proposal tells how someone proposes that a word should be used. The use may be for a special purpose, such as definitions of words such as "drugs" or "poverty" in the stating of laws. Definitional proposals can show a slant aimed at persuading people, such as the definitions of "free enterprise" and "free world." When people stipulate that they are going to use a word in a particular way, they are using a definitional proposal.

One can propose almost any meaning he or she wants for a term. There are, however, some limits to what one can get away with using definitional proposals. Consider the following example and ask whether we can define terms as is done in the example:

Susan: "You're a simple person, Bob."

Bob: "Now wait a minute. I'm smart enough to know when I'm being insulted."

Susan: "I'm complimenting you, Bob. I mean by a 'simple person' one who is in touch with basics."

Bob: "Oh, yeah. Well, thanks."

Sample Test Item

Directions	Indicate what is true of each example using these choices:
	A) It is a definitional report.
	B) It is a definitional proposal.
Example	"A landlord is a leach with property."
Procedure	Ask yourself the following:

1. Does the definition tell how a term is used or how someone wants a term to be used? A report tells how a term is used while a proposal tells how someone wants a term to be used. (In the example, the definition does not tell how the term landlord is used, and is thus a proposal.)

2. Does the definition show the attitude of the person giving the definition? Proposals can be biased, but reports need to be neutral. (In the example, the definition is biased against landlords, and is thus a proposal.)

3. Does the definition report past usage or suggest future usage? Reports tell past usage while proposals suggest future usage. (In the example, this does not apply.)

Note:	Definitional reports, unlike proposals, may be correct or incorrect. If it appears that someone is telling how a term is used it is a definitional report, whether or not the report of usage is correct.

Exercise 2 - 1   **Name** _____

Directions: Indicate what is true of each example using these choices:
A)   It is a definitional report.   B)   It is a definitional proposal.

1.   A liberal is a person who wants to give away another person's money. ....____

2.   A hurricane is a storm with heavy winds rotating around a moving center. ....____

3.   A conservative is a person who wants to advance into the past. ....____

4.   To goof is to make a silly but not costly mistake. ....____

5.   Public servants are officeholders who belong to your party. ....____

6.   Smut -- indecent language or jokes. ....____

7.   Communist -- any liberal New Dealer. ....____

8.   "Quintuplet" means a set of five items. ....____

9.   A prevaricator is a liar. ....____

10. Tutor -- a private teacher. ....____

11. "Terrorist" means a person who uses violence and is not acting on the
     orders of a recognized government. ....____

12. Liberty and justice -- what a candidate begins to talk about when she's
     afraid to discuss her record. ....____

13. "Triangle" means a four-sided closed plain figure. ....____

14. "Triangle" means a small animal commonly called a bed bug. ....____

15. "Triangle" means a three-sided closed plain figure. ....____

16. Bureaucrats are officeholders who belong to the other party. ....____

17. "Quicksand" means an item used in soup to create a beef flavor. ....____

18. A sexagenarian is a person in his or her sixties. ....____

19. A mule is an animal that is half horse and half donkey. ....____

20. A poplar tree is a quick growing tree of the Willow family. ....____

Organizer

A definitional report should correspond to actual usage. For one thing that means that the referent of the definition should be neither broader nor narrower than the referent of the word being defined. We begin with examining the concepts of broad and narrow.

CONCEPTS: Broad and Narrow

Some words have a broader class of referents or refer to more objects than other words. The word "instrument," for example, refers to more objects than the word "medical instrument" since the word "instrument" refers to other instruments besides medical instruments.

Sample Test Item

Directions

Circle the term that is the most narrow.

Example

Instrument, scalpel, medical instrument, artifact

Answer

Scalpel

Procedure

Try the words two at a time. First ask whether the term "instrument' is broader than the term "scalpel." Are instruments scalpels or are scalpels instruments? Which is included under which? The most inclusive term is the broadest. The term "artifact" is the broadest and "scalpel" is the narrowest. Now compare the term "scalpel' with the term "medical instrument." Which is the narrowest? The answer is "scalpel." Finally, compare the term "scalpel" with the term "artifact." Since an artifact is anything that is created by man, and a scalpel is just one object crated by man, the solution is that the term scalpel is the narrowest.

CONCEPTS: Extension and Intension

Definition | Extension - Some terms refer to or comprehend, or denote more objects than other terms. The extensional meaning of a term is what the term denotes or refers to. In the illustration below, you will observe that the term "animal" refers to more than "mammals."

Definition | Intension - The intention of a term includes all the features an object must have in order to be referred to by the term. The intensional meaning of the term "mammal," for example, includes the features of having hair, being an animal, and nursing its young.

Name _____

Exercise 2 - 2

Directions: Answer questions 1 & 2 using the following choices. Then answer questions 3 through 5. Use additional sheets of paper as necessary.

A) Animal   B) Mammal   C) Dog   D) Fido

1. Which term has the largest extension?..____

2. Which term has the largest intension?...____

3. What is the relationship between extension and intension? As we add essential features to an object, what happens to the extension?

    _____

    _____

4. What happens to the number of essential features as we go up the abstraction ladder? Do we know more about something when we hear it is an animal than we do when we hear it is a dog? Consider the two statements, "I bought an animal," and "I bought a dog." Which tells us more?

    _____

    _____

5. Suppose that someone says, "I am for freedom," and someone else says, "I am for free lunch programs." Which person is telling us more?

Exercise 2 - 3

Directions: Indicate what is true of each example using the following choices:
   A)   The underlined word has the broadest extension.
   B)   The underlined word has the narrowest extension
   C)   The underlined word is neither the broadest nor the narrowest.

1. Writing Implements, Implements, electric typewriter, <u>typewriter</u> ....____

2. <u>Geometric Figure,</u> Quadrilateral, Parallelogram, Rectangle, Square ....____

3. The White House, Structure, Residence, <u>Building</u>, Artifact ....____

4. <u>Parisian</u>, European, Westerner, Homo Sapiens, Animal ....____

5. <u>U.S. Wheat for Bangladesh,</u> U.S. Food for Bangladesh, U.S. Foreign
    Aid, U.S. World Leadership. ....____

6. Athlete, Pitcher, <u>Ball Player</u>, Babe Ruth ....____

7. Cat, <u>Wildcat</u>, mammal, Animal ....____

8. <u>Viral infection</u>, Polio, Ailment ....____

9. Violinist, Musician, Artist ....____

10. Apple, Fruit, Winesap, <u>Food</u> ....____

11. Literature, Poem, Publication,"<u>The Raven</u>" ....____

12. Student, <u>College Student</u>, College Freshman ....____

13. Dress, <u>Evening Gown</u>, Clothes, Outer Covering...____

14. Automobile, Chevrolet, Vehicle, <u>Means of Transportation</u> ....____

15. Furniture, <u>Bed</u>, Four-post Bed...____

Organizer - Definitional Inadequacies

These are the types of mistakes a definition report can make:

- Too broad - If a definition denotes more than the term to be defined, then the definition is too broad.

- Too narrow - If a definition denotes less than the term to be defined, then the definition is too narrow.

Sample Test Item

Directions

Indicate what is true of each example using the following choices:
A)   The definition is too broad.
B)   The definition is too narrow.

Example

term to be defined = definition

quart = fraction of a gallon

Procedure

Ask yourself:

1.   Is something included by the definition that should not be included? (In the example pint, ounce, and half gallon are included.) If something is included that should not be included, the definition is too broad.

2.   Ask, is something excluded by the definition that should not be excluded. If so, then the definition is too narrow.

3.   Too broad and too narrow means that at least one thing is included that should not be included and at least one thing is excluded that should not be excluded.

Name _____

Exercise 2 - 4

Directions: Indicate what is true of each example using these choices:
A)   The definition is too broad.
B)   The definition is too narrow.
C)   The definition is both too broad and too narrow.
D)   The definition is neither too broad nor too narrow.

1. Fragrance--any odor ....____

2. Rectangle--geometric figure with four equal sides ....____

3. Rectangle--geometric figure with four right angles............................._____

4. Square--geometric figure with four sides..____

5. Square--geometric figure with four equal sides ....____

6. Amnesia--a mental disorder ....____

7. Misanthrope--a person who hates or distrusts all people ....____

8. Antidote--remedy to counteract the effects of arsenic ....____

9. Painting--picture drawn on canvas with a brush and paint ....____

10. copy--a reproduction ....____

11. Truck--four-wheeled motor driven vehicle ....____

12. Circle--a plane geometric figure whose radii are equal............................____

13. Thief--a shoplifter..____

14. Dog--animal having four legs..____

15. Cassette Player--a device for the reproducing of sounds ....____

Exercise 2 - 5

Directions:

Examine a contemporary problem that interests you and begin by defining words that relate to that problem. An example of this procedure is offered.

Example: Suppose that you are concerned about the possibility that nuclear weapons might annihilate all life, and you would like to work for peace. You might begin by trying to define the term "peace."

Definition 1 - Peace = Disarmament

Ask yourself whether these definitions are adequate. The first definition is too narrow because the goal of disarmament night be achieved without achieving the goal of peace. Nations could rebuild stockpiles of weapons within a few months, so the problem of peace is larger than the issue of disarmament.

Definition 2 - Peace = Ending all Conflicts

The second definition is too broad because peace can be achieved without ending all conflicts. People in the different states of the United States have conflicts with each other, yet they are able to live in peace.

To do this assignment properly, you must start with a tentative definition and then test it by seeing whether it includes something that should not be included, and whether it excludes something that should not be excluded. Then revise your definition and test the revised definition. The point of this assignment is to examine the process of forming and testing definitions. Any of the following words are appropriate for examining: terrorism, freedom fighter, patriot, happiness, free world, socialism, communism, democracy, human rights, love, responsibility, affirmative action, equity and equality. You should revise your definition at least three times, each time explaining why it was either too broad or too narrow.

CONCEPTS: Inconsistent Definition

A definition or set of definitions is inconsistent if inconsistent claims are made within the definition or definition set. Inconsistent claims are claims which cannot be true at the same time. If one claim is true, then the claim that is inconsistent with it must be false.

Remember the concept of inconsistent statements which are statements which cannot be true at the same time. With definitions, it is claimed that objects (the terms to be defined) possess certain properties. If these properties cannot all be consistently possessed by the term to be defined, then the definition of that term is inconsistent.

Example

An Inconsistent Definition:
skeptic = one who knows that one cannot know
                 anything

Procedure

The first part of the definition claims that the skeptic has the property of knowing since he is "one who knows." The second part claims that "one cannot know anything." These are inconsistent claims and so the definition is inconsistent.

Example

An Inconsistent Definition Set:
pond = small lake
lake = body of water other than a river, ocean or pond

Procedure

Assume that the first definition is correct and that a pond is a small lake. Then the second definition is not correct because it says that a lake is not a pond. The first says that a pond is a small lake the second says that a lake is not a pond. Therefore, the two definitions cannot be correct together and are inconsistent.

CONCEPT: Circular Definition

A definition is circular when the term to be defined appears in the definition. In a circular definition, a word is defined in terms of itself. If a person does not know the meaning of a word, a circular definition cannot help because the word that he or she does not know appears in the definition.

	A definition set is a set of definitions which are related in that they share terms. Often when one does not know the meaning of a word used in a definition, one looks for a definition of the unknown word. If the unknown word is defined in terms of the first word the person did not know, then the set of definitions is circular.
Example	A Circular Definition: friend = one who treats you as a friend
Procedure	Note that the word "friend" is defined in terms of itself. If one does not know the meaning of the term "friend," then one does not know the meaning of the expression "treats you as a friend."
Example	A Circular Definition Set: tenterhook = a sharp hooked nail used for fastening cloth on a tenter tenter = a machine or frame for stretching cloth on tenterhooks so that it may dry even and square Note that "tenterhook" is defined in terms of "tenter" and that "tenter" is defined in terms of "tenterhook." If one does not know these terms then the definitions have one going in a circle.

Organizer - Adequate Definitions

Now that we have covered the various types of errors one must try to avoid in formulating definitional reports, we are ready to consider all the difficulties at once. Be careful to avoid confusing one concept with another.

CONCEPT: Adequate Definitional Report

An adequate definitional report is one that does not commit the errors covered. It is a definitional report which . . .
1)   Is not inconsistent
2)   Is not circular
3)   Is neither too broad nor too narrow.

Sample Test Item

Directions   Indicate what is true of each example using the following choices:

A) Circular
B) Inconsistent
C) Too broad
D) Too narrow
E) Adequate

Example   triangle = figure in the shape of a triangle

Answer   A)   Circular

Procedure   1.   Test for each item separately:

a) Is the definition inconsistent?
b) Is the definition circular?
c) Is the definition too broad?
d) Is the definition too narrow?

2. If none of the above apply, then the definition is adequate.

Exercise 2 - 6

Directions: Indicate what is true of each example using these choices:
  A)   Definition is too broad.
  B)   Definition is too narrow.
  C)   Definition is too broad and too narrow
  D   Definition is circular.
  E)   Definition is inconsistent.
  F)   Definition is adequate.

1.   Senior -- a college undergraduate in the fourth year of study. ....____

2.   An overachieving pupil -- one who learns more than it is possible to learn...............____

3.   Desirable -- capable of being desired. ....____

4.   Syllogism -- an argument that is in the form of a syllogism ....____

5.   Automobile -- a machine with a motor. ....____

6.   Painting -- a picture drawn on canvas with a brush. ....____

7.   Fresco -- not a mosaic...____

8.   Novel -- a work of fiction...____

9.   Day is the absence of night and night is the absence of day. ....____

10. Human being -- animals capable of self-consciousness ....____

11. Poison -- anything that can harm an organism. ....____

12. Rectangle -- a quadrilateral with four right angles. ....____

13. Lying -- perjury ....____

14. Ornament -- something that is not necessary for practical use...............................____

15. Piety -- that which distinguishes the pious from the non pious. ....____

Chapter 3:   *FACTUAL DISAGREEMENTS*

Purpose

Arguments are used in the settling of factual disputes. The techniques of logic aid in settling factual disputes, and many other types of disputes can often be rephrased as factual disputes. For example, people may disagree on attitudes because they also disagree on facts. There are tools for settling factual disputes which consist of the methods of argument evaluation. In trying to get people to come to agreement, it is desirable to phrase disagreements, where possible, as factual disagreements and then to evaluate arguments. Understanding arguments is central to this enterprise of coming to agreement.

Instructional Objective

Students will be able to handle factual disagreements with arguments and to classify arguments as inductive or deductive, probable or improbable, valid or invalid, and sound or unsound.

Sample Test Item

Directions

Indicate everything that is true in each example using these choices:

A)	Argument	F)	Invalid
B)	Not an argument	G)	Sound
C)	Deductive	H)	Unsound
D)	Inductive	I)	Probable
E)	Valid	J)	Not Probable

Example

Every American should serve a hitch in the Army. It surely did Bob and Susan a lot of good.

Answer

A, D, J

Behavioral Objective

After completing this chapter, the student should be able to:

1. Define and recognize examples of the following:
   - argument
   - statement
   - premise and conclusion
   - deductive argument
   - inductive argument
   - valid and invalid
   - probable and improbable
   - sound and unsound

2. The student will be able to determine all that is true of an argument.

3. The student will be able to demonstrate an understanding of the concepts covered by criticizing an argument on the basis of either its validity or the truth of a premise.

CONCEPT - Argument

An argument is made of two or more **statements**, one or more of which are the **conclusion** being argued for, and one or more of which are the **premises** or reasons for accepting the conclusion.

Organizer - Identifying Statements

After examining the above definition of an argument, you should note that for something to be an argument it must have:

1. Two or more statements;

2. One or more premises; and,

3. One or more conclusions.

We need to know what a statement is and what premises and conclusions are in order to determine whether there is an argument. The first concept explored is **statement**.

CONCEPT: Statement

A statement is a linguistic unit performing an informative function. Statements state facts. They assert that something is the case. Other words for statement are "proposition" and "assertion." There may be more than one statement in a sentence.

Sample Test Item

Directions

Indicate what is true of each example using the following choices:

A)   It is a statement.
B)   It is not a statement.

Example

"Close the door."

Answer

B

Procedure

Ask yourself the following:

1.   Is the statement true or false?
2.   Does it state a fact?

Exercise 3 - 1

Directions: Indicate what is true of each example using these choices:
  (A)   It is a statement.
  (B)   It is not a statement.

1.   Be sure to sign your tax return before mailing it. ....____

2.   Are you going to the basketball game tonight? ....____

3.   Lorenzo de' Medici lived in the fifteenth century. ....____

4.   Get out of here! ....____

5.   Aha! ....____

6.   Gather ye rosebuds while ye may. ....____

7.   Who was the twenty-third president of the United States? ....____

8.   All humans are homo sapiens. ....____

9.   Light that candle. ....____

10. Kenneth is nineteen years old. ....____

11. Jump in the lake and drown. ....____

12. Relax! ....____

13. Hi there! How are you today? ....____

14. We all know that the judge is honest. ....____

15. No informed person can seriously doubt that the earth revolves
     around the sun. ....____

Organizer - Parts of an Argument

One cannot have a premise without a conclusion, so it is best to define premise and conclusion together.

CONCEPTS: Premise and Conclusion

A conclusion is what a person is trying to prove and the premise or premises are the reasons given to support the conclusion. When one argues, one is supporting facts with other facts. The fact that is supported is the conclusion, while the reasons given for accepting it are the premises.

Sample Test Item

Directions

Indicate what is true in the following example:
A)   The premise is underlined.
B)   The conclusion is underlined.

Example

Mary was accepted at medical school, so she must have had good grades.

Procedure

1.   Key words provide a clue because some words indicate premise, while other words indicate conclusion. Learn the list below.

Premise Indicators	Conclusion Indicators
since	therefore
because	hence
for	so
in as much as	thus
for the reason that	consequently
	it follows that
	we may conclude that

2.   Ask "What supports what?" The premises support the conclusion. If there are two statements, a statement A and a statement B, and B supports A, then B is the premise and A is the conclusion.

3.   See which statement needs to be supported. The conclusion is usually more controversial than the premises. Ask "which statements can stand alone?" to spot the premises, and "Which statements need to be supported?" to spot the conclusion.

Note:	Key words were called a "clue" because these words do not always indicate parts of an argument, nor are they always present. In a sentence "Since the fall of the Roman Empire, Europe has not been united," the word since is not a premise indicator.

The method of looking for the most controversial statement is not always reliable because disagreement on what is controversial can be substantial. |

Organizer - Identifying Arguments

	Now that we know what a statement is and how to tell whether we have premise and conclusion parts, we can tell how to spot an argument. Remember the concept of an argument.

An argument is made up of two or more statements, one or more of which are conclusions being argued for, and one or more of which are the premises or reasons for accepting the conclusion. |
| Sample Test Item | |
| Directions | Indicate what is true of each example using the following choices:

A) It is an argument.
B) It is not an argument. |
| Example | "The sky is blue and the air is warm." |
| Answer | The answer is "B" because one statement does not support the other, and thus there are no premises or conclusions. |
| Procedure | Ask yourself the following:
1. Are there two or more statements?
2. Is there a premise and conclusion? |

Exercise 3 - 2

    Directions: Indicate what is true of each example using these choices:
      A)   A conclusion is underlined.
      B)   A premise is underlined.

1.   Some mammals can fly, since <u>bats can fly.</u> ...._____

2.   <u>Venus and Mercury must revolve around the sun,</u> because of their never moving away from it, and because of their being seen now beyond it and now on this side. ...._____

3.   <u>The defendant is insane.</u> Therefore, he is not guilty. ...._____

4.   <u>The flu is caused by a virus</u>; consequently, it can't be cured with antibiotics. ...._____

5.   <u>Coffee keeps people awake</u>; hence, it must contain a stimulant. ...._____

6.   Since Mr. Scott is a judge, it follows that <u>he is a lawyer.</u> ...._____

7.   <u>That this solution is an acid</u> can be inferred from the fact that it turns litmus paper red. ...._____

8.   This flint knife that we found at our excavation site has to be older than 2500 B.C. because <u>it was found three layers below the layer we dated at 2500 B.C.</u> ...._____

9.   George graduated from Boston University, so <u>he must have lived in Boston.</u> ...._____

10. Lincoln was the ideal leader for a democracy, for he was a man of the people. ......___

11. <u>The city should reimburse Miss Simpson for her hospital expenses</u> for the simple reason that the accident took place while she was doing city business. ...._____

12. A cold is a viral infection and since <u>most medications are ineffectual in treating viral infections</u>, most medications are ineffectual in treating colds. ...._____

13. Because economic growth depends on cheap raw materials and plentiful energy, and these are in short supply, <u>we can expect economic growth to slow.</u> ....___

14. <u>Most people never have to deal with the criminal justice system</u>; thus, they are unaware of how it works. ....___

15. It rained the last three times we planned a picnic. Therefore, <u>if we plan a picnic for Sunday, it will rain.</u> ....____

Name _____

Exercise 3 - 3

Directions: Examine arguments at the back of this book and select one to examine. Fill in what is requested.

Name of the Argument _____

The main conclusion is: _____

_____

_____

_____

_____

_____

One premise offered to support this conclusion is:

_____

_____

_____

_____

Organizer - Argument Types: Deductive vs. Inductive

Once you know that a person is giving an argument, you need to be able to evaluate the argument. In order to evaluate an argument, you need to know the type of argument being offered. There are two main types of arguments -- deductive and inductive -- and each type is evaluated in a different way.

CONCEPTS: Deductive and Inductive

Anyone who offers an argument is claiming that his or her conclusion is true. While all of those offering arguments claim that their conclusions are true, sometimes people claim that their conclusions **must be true**, while at other times people claim that their conclusions are only **probable**. To say that statements must be true is to say that on the basis of the evidence they cannot be false. To say that statements are probable is to say that they are more likely to be true than false. Inductive logic is concerned with probabilities and not certainties.

Whether one is warranted in saying that a conclusion must be true or whether it is only probably true depends on the evidence that one has for one's conclusions. There should be a balance between the evidence and what is claimed on the basis of the evidence. One should not, for example, claim more than one's evidence can support.

The evidence for a claim that a statement must be true can be expected to cover all cases. A deductive statement often has a general statement in it. An example of a deductive argument is: All men are mortal. Socrates is a man. Therefore, Socrates is mortal.

The evidence for a claim that a statement is probable typically cites a number of cases. An inductive argument begins with specific cases and attempts to formulate generalizations. An example is: Rose #1 is red. Rose #2 is red. Rose #3 is red. Therefore, all the roses are probably red.

41

Arguments are considered to be either inductive or deductive. This makes sense when one considers that a person who asserts something to be true is either certain that the statement is true or has reason to think that the statement is probably true. If the claim is considered certain on the basis of the evidence, the argument is a deductive argument; otherwise, it is an inductive argument.

Definitions

In a deductive argument:
- if the premises are true, then the conclusion must be true.
- the premises provide conclusive reason for accepting the conclusion.

In an inductive argument:
- if the evidence for the conclusion is correct, then the conclusion is probable.
- the evidence provides some reason for believing the conclusion.

Sample Test Item

Directions

Indicate what is true of each example using the following choices:
A)   It is an inductive argument.
B)   It is a deductive argument.

Example

"It rained the last three times we planned a picnic and we have planned a picnic for Friday. Therefore, it will probably rain on Friday."

Procedure

To tell whether an argument is inductive or deductive:
1.   Look for indicators:
   - Clues for deduction are such words as "must be," "has to be," "conclusively," etc.
   - Clues for induction are such words as "is probable," "is likely," etc.

2.   Ask whether the conclusion is the sort that one can be certain about. Mathematical examples such as $2 + 2 = 4$ can be certain. Weather predictions, like other predictions of the future, can only be probable.

3. See what effect adding more statements has. Since the premises of a deductive argument provide conclusive reason to accept the conclusion, it makes no difference whether other facts are added to the premises. It makes no sense to speak of more conclusive or less conclusive. It is either conclusive or not. Since the conclusion of a deductive argument follows conclusively from the premises, adding more facts has no effect. Adding facts does affect the probability of an inductive argument.

Since the evidence of an inductive argument provides only some reason for accepting the conclusion, it does make a difference whether other facts are added to the premises. It does make sense to speak of more probable and less probable. Probability is in degrees and adding facts can change the degree of probability.

4. See whether the premises are general or specific -- this rule is not always good, but it does work most of the time.
   - deductive -- goes from general to general or general to specific, uses already made-up generalizations
   - inductive -- goes from specific to general, makes up generalizations

Exercise 3 - 4

Directions: Indicate what is true of each example using these choices:
- A)　It is a deductive argument.
- B)　It is an inductive argument.
- C)　It is not an argument.

1. According to the polls, fifty-four percent of the voters favor Senator Erskine. Therefore, Senator Erskine will probably win the election. ....____

2. All human choices are determined, since all events in the Universe are determined and all human choices are events in the Universe. ....____

3. There are eight people in the class. It follows that at least two of them were born on the same day of the week. ....____

4. Alan Bigwig is the literary agent for Rashish Gibson, Semantha Smith, and Joseph Hanley and all their books sell well. Alan Bigwig just took on Joe Spiro as a client, so Joe Spiro's book will probably sell well. ....____

5. Every art and every inquiry, and similarly every action and pursuit, is thought to aim at some good; and for this reason the good has rightly been declared to be that at which all things aim. (Aristotle, Nicomachaen Ethics) ....____

6. We hear about constitutional rights, free speech and the free press. Every time I hear these words, I say to myself, "that speaker is a Red, that speaker is a Communist." ....____

7. All Merovians are rich. I met two Merovians and they were both rich. ....____

8. Since the base angles of an isosceles triangle are equal, the bisector of the vertex of an isosceles triangle bisects the base. ....____

9. It is likely that prices will rise because energy is used in producing all goods and services, and the price of energy has been rising. ....____

10. Artist are not made great by having great teachers, but are made great by what they themselves do. Raphael, Michelangelo, Hayden, Mozart and most other great masters did not have teachers who were great. ....____

Note: There are several other ways of telling inductive from deductive arguments. These methods include:

1. See whether the argument is based on analysis or observation.
   - deductive = deduce from analysis (analytic)
   - inductive = induce from observation (empirical)
2. Theory vs. practice is the starting place.
   - deductive = theory to practice
     - generalization to instance
   - inductive = practice to theory
     - instance to generalization
3. Understand how each type of argument works.
   - deductive = begins with possibilities, rules out all possibilities except conclusion
   - inductive = outlines range of possibilities, tells probability of answer

CONCEPT: Evaluating Arguments vs. Labeling Arguments

We label arguments according to what the arguments are intended to do. If an argument is intended to have a conclusion that is certain, we say that it is a deductive argument, whether or not the conclusion is in fact certain. If the conclusion is intended to be less than certain, then we say that the argument is inductive.

When evaluating arguments one tests to determine whether an argument does what it is intended to do. Rather than judging what the intention is when one evaluates an argument, one judges whether the argument succeeds in doing what it is intended to do.

Examine the following argument:
> All dogs have four legs.
> All cats have four legs.
> Therefore, all dogs are cats.

Obviously the conclusion that all dogs are cats is not certain. Yet the argument is a deductive argument because the argument is set up as though the person offering the argument intends the conclusion to be certain.

Organizer - Inductive Argument Terms

As mentioned before, there are different terms and tests that apply to different types of arguments. The terms and tests for inductive arguments are now given.

CONCEPT: Probable vs. Not-Probable

Whether one is warranted in saying that a conclusion has a certain probability depends on the evidence that one has for the conclusion. There should be a balance between the evidence and what is claimed on the basis of the evidence. One should not, for example, claim more than one's evidence can support.

Sample Test Item

Directions

Indicate what is true of the following example using these choices:
A) The conclusion is probable.
B) The conclusion is not probable.

Example

It rains every day in Florida. I visited Florida for three days and it rained every day.

Answer

The conclusion is not probable on the basis of the evidence. One cannot conclude that it rains every day in Florida on the basis of three days.

Procedure

Ask yourself:
1. Is the evidence adequate for the conclusion?
2. Is there enough evidence? Is the sample the conclusion is based on large enough?
3. Is the sample on which the conclusion is based representative?

Organizer - Evaluating Arguments

When evaluating arguments we see whether arguments do what they are supposed to do. Deductive arguments are supposed to have conclusions which are **certain** on the basis of their premises, while inductive arguments are supposed to have conclusions which are more likely true than false. Since deductive arguments are supposed to do more than inductive arguments are supposed to do, we have different tests and terms for each type of argument.

CONCEPT: Valid and Invalid

Valid and invalid are terms that apply to a deductive argument. In a valid deductive argument, the premises provide conclusive reason for believing the conclusion; or restated: **In a valid deductive argument, if the premises are true, then the conclusion must be true**. If a deductive argument is not valid, it is invalid.

Sample Test Item

Directions

Indicate what is true of the following examples using these choices:
A) It is valid.
B) It is invalid.

Example

If Susan is seriously ill, then she would miss her class.
Susan missed her class.
Therefore, Susan is seriously ill.

Procedure

Ask yourself the following:
1. Is it possible for the premises to be true and the conclusion to be false? If so, then the argument is invalid. (In the example the premises are true and the conclusion is false and thus the argument is invalid.)

2. If the argument is not invalid by the test above, then it is valid.

Name _____

Exercise 3 - 5

Directions: Indicate what is true of each example using these choices:
- A) Deductive and valid.
- B) Deductive and invalid.
- C) Inductive and probable.
- D) Inductive and improbable.
- E) Not an argument.

1. All those for National Health Insurance are Communists. Mr. Green is for National Health Insurance. Therefore, Mr. Green is a Communist. ....____

2. From all accounts, he is loyal, hardworking, intelligent, cool, able, judicious, and cooperative. Therefore, he will be a success at the job.............____

3. Since no man has natural authority over his fellows, and since might can produce no right, the only foundation left for legitimate authority in human societies is agreement. (*Jean Jacques Rousseau, Social Contract*).....................................____

4. If the sun burnt out then it would be getting colder. It is getting colder. Therefore, the sun must have burnt out. ....____

5. Joe will probably fail the course because he has not opened a book yet. ....____

6. Today is Wednesday. You were here four days ago, so you were here on Saturday. ....____

7. If we take eternity to mean not infinite temporal duration but timelessness, then eternal life belongs to those who live in the present. (*Ludwig Wittgenstein*) ....____

8. Since most of the people from the United States that we see at the market in Toluka, Mexico are wealthy, we may conclude that most American are wealthy......____

9. If Susan Jones were a millionaire, then she could buy a new car. Therefore Susan Jones can buy a new car. ....____

10. Either the tank is empty or the ignition is faulty. The tank is not empty, so the ignition must be faulty. ....____

48

Organizer - Deductive Arguments, Validity and Soundness

When testing deductive arguments for validity, we are testing reasoning. Reasoning has to do with how we put statements together. But reasoning is not enough. One should also have correct facts. When arguments are both valid and have true premises, they are said to be sound. When they are valid and have true premises, they also have true conclusions.

CONCEPT: Sound vs. Unsound

A sound argument is a good argument -- one that people should accept as convincing. In a sound argument the reasoning is good and the statements presented to support the conclusion are true.

Evaluating an argument requires both reasoning and facts. One might ask, "Is the reasoning good?" and "Are the facts correct?"

Sample Test Item

Directions

Indicate what is true of each example using these choices:
A) Sound.
B) Unsound.

Example

All students are eight feet tall. John is a student. John is eight feet tall.

Answer

B. The argument is valid, but because the first premise is false, the argument is unsound.

Procedure

There are two questions to ask when evaluating an argument:
1. Is it valid?
2. Are the premises true?
If both conditions are satisfied, the argument is said to be sound. If either or both are not satisfied, the argument is said to be unsound.

Organizer - Evaluating Arguments

You are now ready to put all the concepts together and to assess whether an argument is being offered, and when an argument is offered to evaluate the argument.

CONCEPT: Assessing Arguments

In assessing evidence for a belief, a person must make several decisions. First, it must be determined whether an argument is being offered. Second, if an argument is being offered, the type of argument being offered must be determined. There are two types of arguments, deductive and inductive, and each is evaluated in a different way. Third, the arguments must be evaluated. Inductive arguments are said to be probable or not probable, and deductive arguments are said to be valid or invalid. Finally, if an argument is deductive, it must be determined whether the argument is sound or unsound.

Sample Test Item
Directions

Indicate everything that is true of each example using these choices:

A)	Not an argument	E)	Valid
B)	Argument	F)	Invalid
C)	Deductive	G)	Sound
D)	Inductive	H)	Unsound

Example

All students are eight feet tall. Bob is a student. Therefore, Bob is eight feet tall.

Procedure

Ask yourself the following:
1. Is it an argument or not an argument?
2. If it is an argument, is it a deductive or inductive argument?
3. If it is an inductive argument, is it probable or improbable?
4. If it is a deductive argument, is it valid or invalid?
5. If it is invalid, then you know it is also unsound.
6. If it is valid, then ask whether the premises are true or false.

Note: Remember:

- **Inductive** and **deductive** are terms that apply to **arguments**.
- **Probable** and **improbable** are terms that apply to **inductive** arguments.
- **Valid** and **invalid** are terms that apply to **deductive** arguments.
- **Sound** and **unsound** are terms that apply to **deductive** arguments.
- The terms **true** and **false** apply only to **statements**.

CONCEPT: Criticizing Arguments

In offering an argument, one makes a claim that the conclusion follows from the premises. The premises are offered as evidence or support for the conclusion. Presenting evidence by stating premises involves two different commitments:

1. A claim is made that the facts listed as premises are evidence for the conclusion.

2. A claim is made that premises are statements of fact, i.e., they are true.

Correspondingly, there are two ways to criticize an argument. First, the premises might not be evidence for the conclusion. They may be unrelated to the conclusion. Second, one or more of the premises might be false and thus the alleged facts may not be facts at all.

In examining an argument, ask yourself:
1. Do the premises support the conclusion?
2. Are the premises true?

Note: If an argument might be criticized either by showing that the premises are not related to the conclusion or by showing that premises are false, it is better to show that the premises are unrelated to the conclusion. This method of criticism is more powerful because, in using it, one can still accept the other person's "facts" and reject his or her conclusion.

51

Exercise 3 - 6

Directions: Criticize the argument below by either pointing out an invalid argument or by pointing out that a premise is false. Indicate what conclusion and premise you are considering. You may use the main conclusion of the entire argument or the conclusion of a sub-argument. To do this assignment correctly you need to write an essay that makes sense by itself, even if the person has not read the following argument.

THE CAUSE OF POVERTY IS OVERPOPULATION

1. The cause of poverty in the world today is overpopulation. 2. There are too many people. 3. We see proof of overpopulation in unemployment, starvation, and the fact that the number of people living in cities worldwide doubles every ten years. 4. It is a fact that the world only has enough resources for a certain number of people.

5. The world's population is growing too large. 6. Holland, the nation with the greatest population density, has two thousand people per square mile. 7. Brazil, Venezuela, and other nations considered overpopulated have considerable poverty. 8. Even the United States must limit its population, as evidenced by the existence of poverty in the United States.

9. It is important for the United States to limit its population. 10. With only five percent of the world's people, the U.S. consumes fifty percent of all goods that are produced. 11. Should the U.S. population double, its people would consume all the world's goods, and if its population tripled, it would be a disaster. 12. All those who are concerned about poverty should take an oath to have only one child, and they should insist that poor nations limit their population growth.

Exercise 3 - 7

Directions: Write an original argument stating your conclusion very carefully in the first line. Write this as though it were an editorial with the heading indicating your conclusion. Everything you state should be relevant to your conclusion.

In grading this paper, your instructor will ask:

1. Is the conclusion clear?

2. Is there support for the conclusion?

3. Is there extraneous material?

4. Does the argument succeed?

Exercise 3-8   **Directions for Writing a Critique of an Argument**

Directions: Write an critique of another person's argument. Your critique should do all of the following:

1. Restate the other person's argument so that your critique would make sense to a person who has not read the original argument.

2. Pick out something essential to criticize.

3. Be clear about whether your are criticizing facts or reasoning.

4. Spot any fallacies that are present.

Chapter 4:   *ARGUMENTS AND FALLACIES*

Instructional Purpose

The three groups of fallacies considered in this chapter concern criteria that any type of argument must satisfy. Remember that an argument consists of two or more statements, one or more of which are premises and one or more of which are conclusions and that premises support the conclusions. All the fallacies in this chapter relate to these facts which are true of all arguments.

Arguments with **language difficulties** fail because sentences that are not clear, or are ambiguous, are not really statements. An ambiguous sentence can be true if you take it one way and false if you interpret it a different way, thus it is not a verbal utterance that is either true or false. The presence of language difficulties violates our definition of argument which states that an argument is made up of two or more statements.

Arguments which have **fallacies of relevance** fail because factors irrelevant to the conclusion are used as premises, thus the premises cannot support the conclusion. Finally, arguments with **self-defeating fallacies** fail because they cannot convince one to accept a conclusion that one does not already accept. For any type of argument to succeed, whether inductive or deductive, the argument must avoid the informal fallacies covered in this chapter.

Instructional Objective

Students will be able to spot informal fallacies.

Sample Test Item

Directions

Indicate everything that is true in each example using these choices:

A)	*Ad hominem*	D)	Force
B)	*Ad populum*	E)	Pity
C)	Authority	F)	Ignorance

Example

"It is, too, my turn to pitch. After all, it's my ball."

Answer

D) Force

Behavioral Objective   After completing this chapter, students will be able to spot any of these fallacies which may be present in an argument:

- Language Difficulties
    - No Clear Meaning
        - vague
        - obscure
        - excessive abstraction
        - jargon
    - Two Meanings
        - semantical ambiguity
        - syntactical ambiguity
        - equivocation
- Fallacies of Relevance
    - *ad hominem*
    - authority
    - *ad populum*
    - force
    - pity
    - ignorance
- Fallacies of Self-Defeat
    - inconsistent
    - circular
    - irrelevant conclusion

Organizer - Fallacies Involving Diction

Before one can evaluate an argument one must be able to interpret the argument; one must be able to know what it means for the premise or the conclusion to be true or false. When language difficulties are present, it is not possible to know what the statements in an argument mean.

CONCEPT: Two Kinds of Language Difficulties

Language difficulties are of two sorts. One might not be able to determine the meaning of a sentence because the sentence has two or more meanings, or one might have difficulty because one cannot find even one clear meaning. These difficulties might be labeled as follows:

- No clear meaning
- Ambiguity - two meanings

56

Procedure	To test for the presence of language difficulties, ask yourself :

1. Can you tell what the statements mean?
2. Is there at least one clear meaning? If there is not even one clear meaning, then there is a fallacy of clarity.
3. Can the statements be taken two ways? If so, then there is a fallacy of ambiguity.

Organizer - Fallacies of Clarity

The type of language difficulties to be discussed are the ones that involve no clear meaning. After that, the cases where there are two meanings will be discussed.

Four fallacies that involve no clear meaning are vague, obscure, excessive abstraction, and jargon.

CONCEPT: Vague

Definition	Vague -- An expression that is more indefinite than it should be. It is always best to be as definite as one can be.
Example	To say that "someone called" when you know Bob called.
Procedure	Ask, "Is the expression as definite as it can be?" Note that the concept of being definite is relative. The idea is that one should be as definite as one can be.

CONCEPT: Obscure

Definition	An expression is obscure when something is said in a way that is much more complex than necessary. There is meaning, but it is hard to find and once it is found, you can tell that the expression could be stated more simply.
Example	"He is not not going."
Procedure	Ask:

1. Is there a meaning?
2. Could the expression be put more simply?
3. Can you think of a way of expressing it more clearly?

CONCEPT: Excessive Abstraction

Definition

Excessive abstraction occurs when a person connects one abstraction with another so that it is not possible to tell what they are talking about. Ideally one should go from one level of abstraction to another. With excessive abstraction, the person is stuck on one high level of abstraction.

Example

"I am for freedom and democracy," when asked if he supports abortion laws.

Procedure

Ask yourself:
1.   What is the person's view?
2.   If you cannot tell what the person's view is, is the problem due to using abstract terms?

CONCEPT: Jargon

Definition

Jargon is present when specialized terminology is used in instances when ordinary words would suffice. While the use of specialized vocabulary is sometimes necessary, specialized vocabulary may impede communication in many contexts.

Example

He has an acute inflammation of the mucous membranes of the trachea passages. (He has a cold.)

Procedure

Ask yourself: Does the use of specialized terms aid communication or hinder communication? If it hinders communication then jargon is being used.

Exercise 4 - 1

Directions: Spot the fallacies in the following passages using these choices:
A) Vague   C) Excessive Abstraction
B) Obscure   D) Jargon

1. After conducting an examination, Dr. Barnes decided that the patient had pneumonia. When the patient asked Dr. Barnes what he had, the doctor replied, "You've caught something." The patient's wife then asked how serious it might be. The doctor replied, "It's not true that he won't be going to the hospital." When pressed for an explanation of what was wrong, Dr. Barnes replied, "He's suffering from an inflammation of the trachea." When the patient in desperation asked for the truth from Dr. Barnes, Dr. Barnes gave it to him. He said, "For what are we but specks in the universe, and what is this ailment, but a brief moment in the history of creation."

2. Mr. Metafiz took his place at the podium and began his speech. "I am, you are, and so is he," Mr. Metafiz began. "Since we all are, we all share being," he continued. "Being is a creature itself although it has no features and cannot be described. Opposed to Being in an endless fight is Non-being. Non-being is also a thing in itself and we are constantly caught in the tension between Being and Non-being. Indeed, the struggle between the two goes on in all of us."

3. "I am for freedom," proclaimed the political candidate.

    "Are you for free school lunches?" asked a voter.

    "I am for the kind of freedom that is compatible with Democracy," was the reply.

    "Are you for allowing people the right to settle issues in state referendums?"

    "Provided that the referendums serve the public interest."

Exercise 4 - 2

Directions: Rewrite the following sentences avoiding expressions that are vague, obscure, excessively abstract, or use jargon. Avoid language which is pretentious, or uses cliches or slang. Whenever it is possible to cross out sections of a sentence without affecting its meaning, it is best to do so.

1.   The hoofers pirouetted across the ballroom, but they had to cease and
     desist on the stroke of midnight.

2.   Her anorexia made her thin as a rail.

3.   In my opinion, the reason why she holds that view is because she is selfish.

4.   The automobile rolled backwards out of the driveway and devastated a tree.

5.   The preacher laid his message on the line.

6.   My vast expenditures at the grocery store left me with very little cash flow.

7.   Your eldest brother who is the eldest one in your family speaks in a very
     loud tone of voice.

8.   I do not know what to do at the present time.

9.   The reason why he did that is that he didn't want to not do anything.

10. All patriotic Americans support keeping America strong.

CONCEPT: Ambiguity - Semantical and Syntactical

If a sentence can be taken two ways, then there is an ambiguity in the sentence. The sentence is said to be ambiguous.

If a sentence has two meanings it is because something can be taken two ways: either a word in the sentence or the structure of the sentence. Remember that language is made up of **words** that are joined **grammatically**. If there is a difficulty, it is due either to the words that are used or to the way the words are put together. When one speaks of words they are speaking of semantics and when one speaks of grammar they are speaking of syntactics. Thus, we may speak of semantical and syntactical ambiguity.

Definition of Semantical Ambiguity

When a sentence can be interpreted two ways due to a word being taken two different ways the sentence is semantically ambiguous.

Example

No one cared for the little puppy.
This could be taken to mean:
1.   No one took care of the little puppy.
2.   No one loved the little puppy.
The difficulty in this case is that the phrase "cared for" is ambiguous.

Definition of Syntactical Ambiguity

When a statement has two meanings due to there being two ways of interpreting the grammatical structure of the sentence, the statement is syntactically ambiguous.

Example

A story is told about the Oracle at Delphi who was supposedly able to predict the future. The Greeks were going to battle the Persians and consulted the Oracle, who said, "The Greeks the Persians will conquer." His answer could be taken to mean:

1.   The Greeks   the Persians will conquer
        subject   *direct object   verb*
2.   The Greeks   the Persians will conquer
        direct object   *subject   verb*
In the example, the word "Greeks" is taken as the subject of the sentence in #1 and as the direct object in

#2. It is syntactically ambiguous because the grammar can be taken two ways.

Procedure

1. First, test for ambiguity by asking " are there two meanings?

2. Is only one meaning intended? With ambiguities only one meaning is intended. The problem is that we cannot tell which meaning to use.

3. Determine how the two meanings differ.
   a. Do they differ due to a word -- if so then the problem is semantics.
   b. Do they differ due to grammar -- if so then the problem is syntactical.

4. Ask yourself, how would you correct the sentence so there is one meaning?

5. Test for grammar problems; see if a word switches grammatical function.

Organizer - Using Two Meanings

With an ambiguity only one meaning is intended and only one meaning is used but the reader has no way of telling which of the two meanings to use. A related fallacy involves have two meanings and using two meanings. The difficulty is that those who use this fallacy are not aware that the word has two meanings.

CONCEPT: Equivocation

An equivocation occurs when a switch is made from one meaning to another in a context that calls for only one meaning. One context that calls for only one meaning for terms is an argument. The premises of an argument should discuss the same subject as the conclusion. When a switch of topics occurs, an equivocation is present. Another context in which only one meaning is called for is in conversation. When people are talking and using the same words, they should also be using the same meanings for those words.

Example	There are laws of nature, and where there are laws there must be a lawmaker, so there must be a God.
Procedure	Ask yourself: 1. Are there two meanings of a word? (In the example, there are two meanings of the 'laws.') 2. Are both meanings being used? (In the example, both meanings are used.) 3. Is there a switch from one meaning to another? (In the example, there is a switch from 'laws' meaning regularity to 'laws' meaning bills passed by a legislature.)
Note:	Your understanding of logic will aid, and be aided by, your understanding of grammar. Often when a sentence is syntactically ambiguous, it is because there is a "dangling modifier," which means that the reader cannot tell what is being modified. To avoid ambiguity, modifiers should be placed near the words they describe. The following sentence is ambiguous because it is not clear what the phrase *"walking down the street"* is modifying: *"Walking down the street, the gardens looked lovely."* The ambiguity is removed when the phrase *"walking down the street"* is placed next to the understood subject, "I", which is now stated to avoid ambiguity: *"Walking down the street, I saw that the gardens looked lovely."*
Note:	If you think that the answer must be either semantical or syntactical ambiguity, select syntactical ambiguity as your choice. Quite often an ambiguity in the meaning of the word only arises because the grammatical structure is taken two ways.

Exercise 4 - 3

Directions: Indicate what is true of each example using these choices:
A)   Semantically Ambiguous   C)   Equivocation
B)   Syntactically Ambiguous   D)   None of the Above

1.   For sale: Antique table suitable for gentleman with wooden legs. ....____

2.   How do you explain the fact that you are responsible for your acts when you are not responsible for them? ....____

3.   He looked at the burglar, his eyes wide with fear. .... ____

4.   Nothing is too good for you. ....____

5.   After the general watched the lion perform, he was taken to the city hall and fed twenty-five pounds of meat. ....____

6.   I shall lose no time in reading your research paper. ....____

7.   Out of gas she had to walk home. ....____

8.   The gorilla is more like a man than a chimpanzee. ....____

9.   I went out onto the front porch and watched the fireworks go up in my pajamas ...____

10.   While we were eating a young man the son of the proprietor came in. .... ____

11.   It is man's nature to categorize and discriminate. Since it is man's nature to discriminate, there will always be discrimination by one group against another group. ....____

12.   Her father has a very distinguished appearance, so he must be a very distinguished man. ....____

13.   You can be sure that I am behind you, yes, far behind. .... ____

14.   A crumb is better than nothing. Nothing is better than strawberry shortcake. Therefore, a crumb is better than strawberry shortcake. ....____

15.   Joe loved alcohol better than his wife. ....____

Exercise 4 - 4

Directions: Rewrite those sentences in the previous exercise that were syntactically ambiguous. Be sure that parts of speech -- noun, verb, direct object, and modifiers -- are clear and that modifiers are placed correctly so that no ambiguity remains.

_____

_____

_____

_____

_____

_____

_____

_____

_____

_____

_____

_____

_____

_____

_____

_____

_____

Organizer - Fallacies of Relevance

In addition to having two statements, an argument must have a premise and a conclusion. The premises must be related to the conclusion in order for the premises to support the conclusion. When fallacies of relevance are present, the premises are not relevant to the conclusion.

CONCEPT: Relevance

Fallacies of relevance are present when the truth of the premises is unrelated to the truth of the conclusion. In an acceptable argument, the premises provide a reason for believing the conclusion. When an argument is acceptable the truth or falsity of the premises is related to the truth or falsity of the conclusion.

Sample Test Item

Example

"Smith should not be elected dog catcher. After all, he's a divorced man."

Procedure

Ask yourself, "Does it matter whether the premises are true or false?" In this case, "Does it matter whether Smith is divorced or not?" Does the truth of the premise relate to the truth of the conclusion? Are the premises really evidence for the conclusion?

In the example, the mentioning of Smith being divorced is irrelevant to the conclusion of whether or not Smith would make a good dog catcher.

Organizer - Other Factors Used Fallaciously in Arguments

There are various factors which are usually not relevant but which are often used fallaciously in arguments. These factors and the names of the fallacies are given in this section. Remember the paradigm situation for using arguments. People in a situation of disagreement should try to convince each other by using facts to support their conclusions, and they should consider each other's arguments. When fallacies of relevance are present, people do not direct their attention to what others are saying, but instead to irrelevant factors, such as: who is speaking; who else supports a view; or the situation of a person who supports a view.

CONCEPT: *Ad Hominem* **and Mis-placed Authority**

It is irrelevant to mention who the person is who offers an argument.

Ad hominem -- an argument directed at a person. This argument works by mentioning who the person is who suggests a view, or something about a person, and argues for or against a conclusion on the **basis of the person** suggesting it.

Authority -- This argument argues for a view on the **basis of naming an authority** who supports it. If a person who is an "expert" in one area is cited as the authority to be listened to in an area which is irrelevant to the person's expertise then there is a fallacy of mis-placed authority. The other variant of this fallacy occurs when the concept of authority is overused, such as to claim that something is true because it appears in a book. The term "authority" includes both mis-placed authority and the overuse of authority.

Procedure for Spotting *Ad Hominem* and Mis-Placed Authority

1. Both *Ad Hominem* and Mis-placed Authority mention who supports the conclusion, or something about the person who supports the conclusion, and this is irrelevant.

2. If the argument rejects a conclusion because of the person suggesting the conclusion, or something about the person, then it commits the *ad hominem* fallacy. Authority always argues for the acceptance of a speaker's view whereas *ad hominem* may be positive or negative.

3. If the argument accepts a conclusion because of the testimony of a person or organization which is named as supporting the conclusion, then the fallacy of authority may be present. An authority must always be named for there to be a fallacy of authority.

67

CONCEPT: *Ad Populum* Fallacies

It is irrelevant to mention the number of people or the type of people who support a view. The fact that a great many people support of view does not insure that truth of that view. Also to point out that a certain type of person supports a view does not insure the truth of that view. There are several forms of the *ad Populum* fallacy. First, there are instances in which the **number of people** supporting a view is mentioned. Second, there are instances in which the **type of people** supporting a view is mentioned, and third, there are **low level emotional appeals**.

An *ad populum* fallacy that involves mentioning the number of people who support a view is the bandwagon fallacy. This involves an appeal to do something or accept a view because everyone else is joining in this. This kind of reasoning does "Do it because others are doing it." It must be reasonable because everyone says it is reasonable. And in arguments, "the conclusion must be true because this is accepted by most people."

A second variety of the *ad populum* fallacy mentions the type of person who supports a view. This involves the emotional identification with groups of people. A car may be sold by pointing out that many rich people buy the car. This is a fallacy because the fact that many rich people buy the car does not show that it is reasonable to buy it.

A third group of *ad populum* fallacies are related to the meaning of the word "populum" in Latin. *Ad populum* in Latin means literally an appeal to the masses. In Rome, "populum" referred to lower class people, and so an *ad populum* appeal was one to the lower sorts rather than to the educated aristocrats. The fallacy "appeal to the gallery" has the same meaning since the poor sat in the gallery. Any low down appeal based on the emotions, such as stirring up fear or hatred, may involve the *ad populum* fallacy.

Example

"You should subscribe to the book of the month club. Most of your town leaders have already subscribed."

Procedure	1.	Does the argument mention the number of people who support a view? If so, then the argument involves an *ad populum* fallacy.
	2.	Next, examine the argument. Is the argument based on naming one person or a group of people. (In the example, it is based on a number of people -- a group -- who subscribed, so the fallacy is not *ad hominem* and must be *ad populum*.)
	3.	Also ask, "Does the argument involve an emotional identification with a group or image?" (The example does involve the identification with "town leaders.") If so, the argument involves an *ad populum* fallacy.

Procedure to Separate *Ad Hominem* and *Ad Populum* Fallacies

1. First, test for whether the argument involves an irrelevancy. (Referring to the above example, would it matter whether or not most leaders had subscribed? It would not, so there is an irrelevant premise.)

2. Next, examine the argument to determine whether the argument is based on naming one person or a group of people. (It is based on a group of people who subscribed, so the fallacy is not *ad hominem* and must be *ad populum*.)

3. Also ask, "Does the argument involve an emotional identification with a group or an image?" (It does involve the identification with "town leaders.") If so, the argument involves an *ad populum* fallacy.

CONCEPT: Situation Fallacies

What a person's situation is is irrelevant to the truth of their statements. Some situations which people find themselves in which are irrelevant are:
- force -- when a person is in a position of power over another person
- pity -- when a person is at another person's mercy
- ignorance -- when a person, due to their own situation, does not know something

69

Procedure	Does the person claim that their conclusion is true due to their situation?

CONCEPT: Force

The fallacy of force is used when one mentions or hints at consequences to be suffered by the person the speaker is trying to convince if a view is not accepted. The person "being convinced" may feel fear.

CONCEPT: Pity

The fallacy of pity is used when one mentions or hints at consequences to be suffered by someone other than the person the speaker is trying to convince if a view is not accepted. The person "being convinced" may feel sympathy.

Sample Test Item

Example	"It is too my turn to pitch. After all, it's my ball."
Procedure	Ask yourself: 1. Are the consequences mentioned or hinted at? (The person hints that they will take their ball and go home.)
	2. Will the consequences be suffered by the person the speaker is trying to convince? If so, the fallacy is force. If by someone else, then the fallacy is pity. (In the example, the ones he is trying to convince will suffer if he takes the ball home.)

CONCEPT: Fallacy of Ignorance

The fallacy of ignorance is present if one uses ignorance as evidence. An appeal to ignorance usually goes, "I don't know, therefore ...."

Sample Test Item

Example	"Nixon's first trip to China must have been a waste because I never heard of any agreements that were made."
Procedure	Ask, "Is ignorance being used as evidence?"

Directions: Indicate what is true of each example using these choices:
      A)   Ad Hominem   C)   Ad Populum   E)   Pity
      B)   Authority   D)   Force   F)   Ignorance

1.   We must conclude that they have extrasensory perception, since no
      one has ever proven they do not have this ability. ....____

2.   Student to teacher: "the statement is true because I read it in the
      textbook"...____

3.   This book is no good. Its author was an atheist. ....____

4.   The earth is flat. This is true because we find the statement in the
      poetry of Homer. ....____

5.   He will make a good class president. Not only is he a star football
      player, but he is a crew man as well. ....____

6.   Join the switch to L&M cigarettes. ....____

7.   The archbishop of Canterbury calls this the finest art exhibit ever
      assembled, so it must be the finest art exhibit..................................____

8.   Of course, class attendance is important. If you don't attend your
      classes regularly, you will be placed on academic probation. ....____

9.   The minister had just finished denouncing vices prevalent in the
      community. One of the members of the congregation said to another,
      "Don't pay any attention to her. She doesn't practice what she
      preaches." ....____

10. I do not think the failure you gave me on that last examination is
      justified. Besides taking care of my three children when my wife works,
      I have to work the swing shift at the factory. ....____

11. General McWilliam, one of the world's greatest military geniuses, says that
      we should vote for John Clements. That is the clincher. At the next election,
      John Clements gets my vote. ....____

12. She may be a well-known child psychologist but as far as I am concerned
      the fact that she has no children of her own tells me a great deal about her
      advice on how to raise children. ....____

13. Teacher to student: "If you don't accept my reason for marking your answer wrong, I shall regrade your paper; but you can be sure that your score will not be as high as it is now." ....____

14. How can you accept his recommendation that surgery be performed? After all, Dr. Blank is a surgeon and surgeons are expected to recommend operations. ....____

15. Councilman Smith's plea that the city needs more industry cannot be accepted as worthy of serious consideration, since she is a member of the Chamber of Commerce and any member of the Chamber of Commerce is expected to make this kind of plea. ....____

16. If you don't agree that I deserve to be reelected to Congress, then I will see to it that no new highways are built in your area. ....____

17. Opposing military training on this campus is the unenviable privilege of a few misfits and pacifists. ....____

18. We have no evidence to show that Mr. Ferris is not a spy. Let's arrest him. ....____

19. The trouble with your idea is that you are too young to be proposing it. ....____

20. Come on and try it. Everyone does it these days, so it can't be wrong. ....____

In order for any argument to succeed, it must have separate premise and conclusion parts, must have premises that can be true, and must have a conclusion that is related to the premises. In fallacies of self defeat, the purpose of argumentation is defeated because these conditions are not satisfied.

CONCEPT: Inconsistent Premises

When the fallacy of inconsistent premises is present not all the premises can be true at the same time. An argument with inconsistent premises defeats itself because it can never be sound.

Procedure

Ask yourself, "Can all the premises be true at once?" Imagine that the first premise is true. If it is true then can the second premise also be true? Try this for all the premises. If all the premises cannot be true at once then the fallacy of inconsistent premises is present.

CONCEPT: Circular Argument (Begging the Question)

In a circular argument premises are not really separate from the conclusion but instead the premises assume the truth of the conclusion. The fallacy of circular argument is also called "begging the question" because a person who does not already accept the conclusion cannot be convinced by an argument that uses the conclusion as a reason for accepting the conclusion.

Procedure

You might use any one of these methods:
1. Examine the argument to see if the premises restate the conclusion.

2. Examine the conclusion and see whether after considering the argument a person might have any more reason for accepting the conclusion than they had before considering the argument.

3. You usually feel that a circular argument goes in a circle. In reading the argument you feel that you are right back where you began.

73

CONCEPT: Irrelevant Conclusion (*Non sequitur*)

When an argument has the fallacy of irrelevant conclusion the premises are not related to the conclusion. Typically the conclusion brings up a new topic which is not mentioned in the premises. The Latin name for the fallacy -- *non sequitur* -- is instructive in that it means "does not follow." When listening to a person who uses a *non sequitur*, the listener gets the feeling that there is no sequence to the thoughts.

Procedure

Ask yourself:
1. Does the conclusion bring in a topic that is not mentioned in the premises?
2. Can you sense whether there is a gap in thought which shows no sequence? Typically an argument with a non sequitur will begin by making sense and then have a gap in the reasoning when the new topic is introduced in the conclusion.

Note:

With both circular arguments and irrelevant conclusions your best guide is to trust your feelings. With circular arguments you will feel that the argument is going in a circle. With irrelevant conclusions you will feel that there is a gap in the argument.

Name _____

        Directions: Indicate what is true of each example using these choices:
               A)   Inconsistent Premises
               B)   Circular Argument
               C)   Irrelevant Conclusion
               D)   No Problem

1.   The new student says I am his favorite professor. And he must be
     telling the truth, because no student would lie to his favorite professor. .... ____

2.   Englishmen wore wigs in the sixteenth century. Therefore, Shakespeare
     was a great dramatist. ....____

3.   A phone company has advertised that one phone is not enough in a
     modern home because a modern home is one with many phones ....____

4.   It doesn't make any difference whether you win or lose a war. Everyone
     loses in war.   ....____

5.   We believe in the truth of the Bible because it is the word of God, and we
     believe that it is the word of God because in the Bible it says that the Bible
     is the word of God. ....____

6.   All people who like the novels of Emily Bronte are persons with
     excellent literary taste, since persons with excellent literary taste are
     persons who like Emily Bronte's novels. ....____

7.   It is necessary to confine criminals and dangerous lunatics. Therefore there
     is nothing wrong with depriving people of their liberties ....____

8.   Why should I do this?
     Because it is right.
     Why is it right?
     Because God commands it.
     Why does God command it?
     Because it is right. ....____

9. The constitution guarantees free speech. Therefore, if a person believes
     that the only way to achieve certain reforms is through rioting, they should
     have the freedom to riot. ....____

10. John says that he loves me and he must be telling the truth, because
     a person who says that he loves someone would never lie to the person
     he loves. ....____

11. Every person has a right to promote and advance his or her ideas. Therefore, judges and other public officials are justified in using their official positions to advance their religious views. ....____

12. This problem is wrong. It's wrong because it has a mistake in it. There is a mistake in it because it's incorrect and it's incorrect because it's wrong. ....____

13. "I only go to good movies." "How do you know that they are good?" "Well, I don't go to them unless they're good." ....____

14. London, England, is an interesting and beautiful place to visit. But there is no city in Great Britain that is worth your time, at least in terms of a visit, because most are ugly and the rest are totally uninteresting ....____

15. Any great musical composition is beautiful because if it isn't beautiful then it is not a great musical composition. ....____

16. If the gods cause men to be sick, sickness must be good, and its opposite health, must be evil. Likewise, if the gods cause men to be healthy, health must be good, for all that the gods cause must be good. ....____

17. Sil: The strongest motive always determines a person's choice.
    Sal: But how do you know the strongest motive is?
    Sil: You can determine the strongest motive by the choice that is made. ....____

18. Jim is a college graduate. So he must be able to drive a car. .... ____

19. The Governor must be a good friend to the farmers of this state, because he told them so in his speech last night, and no one would lie to his friends. ....____

20. If Jessica wants to go to the movies, then Susan will want to go with her. Jessica wants to go to the movies. So Andrea will not be going to the movies. ...._____

Chapter 5:   *CONFLICTS OF INTEREST*

Instructional Purpose - Introduction to Conflict Resolution

> People have always been interested in learning ways of resolving conflicts. Even those people who want to go to war with others prefer to have those on their side settle their conflicts without violence. It is only recently, however, that the study of conflict resolution has begun in earnest. In this chapter some of the vocabulary, concepts and applications of the new discipline of conflict resolution are presented.

Organizer - Responses to Conflict

> There are various responses that can be taken to the existence of conflict. Several of these are presented.

CONCEPTS: Conflict Avoidance, Negotiation, Arbitration, and Mediation

> Conflict Avoidance occurs when a party to a conflict decides to forgo his or her interest in favor of another party. Conflict avoidance is also called capitulation. People choose this strategy when they decide that trying to pursue their interest is not worth the trouble.

> Negotiation occurs when disputants try to reach an agreement that will satisfy the basic self-interest of each party. Negotiation depends on the ability to make promises that can be kept.

> Arbitration occurs when parties to a dispute either decide to or are forced to accept the decision of a third party. The third party is called an arbitrator or arbiter.

> Mediation occurs when disputants decide their own process and decision settlement. A mediator helps the parties develop options and meets separately with the parties to help them see options realistically so that they can come to an agreement that all parties accept. Mediation is a voluntary process.

Organizer - Mediation

> We will examine mediation because that is the peacemaking role that individuals can hope to play. A mediator helps to bring disputants together. Certain skills aid the mediator in this process. One skill is to have the disputants focus on interests.

Organizer - Conflict of Interest

> A conflict of interest occurs when two or more parties want something -- the attainment of which by one precludes the attainment by the other. Conflicts are real or perceived states of competing interests. People may perceive that they have a conflict when do not actually have competing interests.

CONCEPTS: Interests versus Positions

> Interests are end results such as the satisfaction of needs.

> Positions are views people hold regarding the means they might use to achieve ends.

> The difference between interests and positions is the same as between ends and means. It is best to focus on interests and not positions because there may be more than one means to an end

CONCEPT: Person Centered, Position Centered, Interest Centered

> Person centered disputes exist because of the personalities involved.

> Position centered disputes exist because people are taking incompatible positions. Another name for position centered disputes is "issue centered." Positions are statements regarding means for the achievement of ends.

> Interest centered disputes exist because people have conflicting interests. If one person gets what he or she wants then the other parties cannot get all of what they want. The emphasis is on end results and the satisfaction of needs.

78

Exercise 5 - 1

Directions: Indicate what is true of each example using these choices:
- A) This is centered on personalities.
- B) This is centered on positions.
- C) This is centered on interests.

1. The union would like to discuss health benefits and wage increases. ....____

2. We must stick together as a people if we want to make progress. ....____

3. Neither the United States nor the Russia would benefit from the spread of nuclear weapons to more countries. ....____

4. Let us begin the mediation session by examining the statements made by each party in the dispute. ....____

5. We will begin this mediation session with each person stating what they think of the other person. ....____

6. I would like to begin this mediation session by asking each person to say what bothers them about the existing situation. ....____

7. I don't see how Kate can talk to that guy. ....____

8. You saw our list of demands. That is all we are here to discuss. ....____

9. Mark does not see how Joe can put up with Sue's flirting with others..................____

10. Neither the Jones nor the Smiths will benefit from the fighting. . ...____

11. My fellow brothers in the fraternity and I will stand firm on our decision. ....____

12. I will never pass this class as long as Dr. Brooks is the instructor. ....____

13. We will begin this mediation session with each person stating what they think the decision should be. ....____

14. Our viewpoint is clearly stated in our sheet outlining what we think the final settlement should contain. ....____

15. Let me tell you how I am hurt by what has happened. ....____

It is best to get the parties to a dispute to discuss their interests, rather than personalities or positions, because there are techniques for reconciling interests. Some of the possible outcomes in terms of interests are presented.

CONCEPTS: Competition, Compromise, Collaboration & Cooperation

Competition is a strategy that a person in conflict may follow which would, if successful, result in the person getting all of what they want, with the other parties to the conflict getting nothing.

Compromise is a strategy that aims at give and take so the result is 50-50.

Collaboration is a strategy that involves the integration of interests so that a disputant helps the party they have disputes with get what they want while still protecting their own interest.

Cooperation is working together for mutual gain and sharing the gain equally.

Exercise 5 - 2

Directions: Indicate what is true of each example using these choices:
  A) Example of competition   C) Example of collaboration
  B) Example of compromise   D) Example of cooperation

1. If I help him get what he wants, maybe he will help me get what I want. ....____

2. I don't want to be frustrated, so it's too bad for her. ....____

3. She and I can work together to achieve our common goal. ....____

4. I'll grab the biggest piece of the cake before he does. ....____

5. I'll get to know her better and try to help her realize her dreams. ....____

6. I tampered with her racing car, so I am sure to beat her. ....____

7. If I help you with your English, will you help me with my math? ....____

8. I want to get to class before all the seats are taken. ....____

9. Since I don't want her to get a better grade than I get, I won't let her know the assignment. ....____

10. He and I can study together so we can both pass the course. ....____

11. I watched my show for thirty minutes and he watched the basketball game for thirty minutes. ....____

12. We want a twenty percent salary increase and don't care what it does to the company. ....____

13. We will meet you half way on our demands for a twenty percent salary increase. ....____

14. We would be willing to take a salary increase in the form of a stock-sharing program if this would help the company. ....____

15. You heard my offer. Take it or leave it. ....____

The mediator's job of helping the disputants come to agreement is best aided when all sides are aiming at a solution where everyone wins. The ideal is to have collaboration, but compromise may be the best solution possible. The main tool that the mediator has is the ability to ask the parties to the conflict questions that will help them be realistic about what the other side might agree to. The next section discusses the different types of questions a mediator might ask.

CONCEPT: Open Questions vs. Closed Questions

Procedure	Ask yourself, "Is the answer yes or no?" If so, it is a closed question. If the range of answers is open then it is an open question.
Examples	"What do you hope to get from this process?" (open) "Are you here to try to reach an agreement?" (closed)

CONCEPT: Direct vs. Indirect

Procedure	Ask, "Do they have to answer?" If yes, it is a direct question. If no, it is an indirect question.
Examples	"What do you propose?" (direct) "I wonder why she is resisting your offer?" (indirect)
Note:	It is possible to combine the distinction between open and closed with the distinction between direct and indirect. The four possibilities, and examples of each, are:

Open Direct	Why do you think he will do it?
Open Indirect	I wonder what he will do?
Closed Direct	Do you think he will accept?
Closed Indirec	I wonder if he will accept that?

Exercise 5 - 3

Directions: Indicate what is true of each example using these choices:
A) Open Direct   C) Closed Direct
B) Open Indirect   D) Closed Indirect

1. I wonder why she hasn't suggested a settlement along these lines? ....____

2. Do you think that your proposal will be accepted? ....____

3. Would you be willing to state this directly to her? ....____

4. What do you think his response will be to your proposal? ....____

5. I wonder if she will accept your offer? ....____

6. I wonder why Bob chose to go with Plan B? ....____

7. Why do you think that Bob decided to go with Plan B? ....____

8. I wonder what he will do at the next meeting? ....____

9. Will you talk to him? ....____

10. Do you feel that they will do as you suggest? ....____

11. Do you think she will accept your offer? ....____

12. I wonder if you would be willing to say that to her directly? ....____

13. Do you think he will mind? ....____

14. What do you think is her reason for being there? ....____

15. I wonder what she might think? ....____

CONCEPT: Objective Criteria

Objective criteria are criteria that are independent of individual's will. When settlements are not based on objective criteria, people pit their will against each other. The fault with this is that one party ends up the winner and the other the loser, and no one wants to be the loser. It is desirable to have some basis for settlement that is independent of will, to base decisions on principles that all parties can accept.

CONCEPT: Fair Standards

Fair standards are standards for determining a settlement that either party might accept, even if they were in the other party's position. There can be more than one objective standard. Consider this example: A house has burnt down and the owner has a dispute with the insurance company over the price. Fair standards for the house value might be the replacement cost, the cost of building the house, the market value, or what a court would probably award. If the standard is fair then it is one that the insurance adjuster and the home owner would accept even if they were in each other's position.

CONCEPT: Fair Procedures

If procedures are fair, then either party in the dispute would be willing to take the other party's position. An example of this is the tried and proven method of dividing a piece of cake. One cuts and the other chooses. It doesn't matter very much which person does the cutting and which does the choosing.

Note:

The use of fair standards and fair procedures introduces concepts of morality into conflict resolution. Indeed, appealing to moral principles does help the resolution of conflict. Conflict resolution is greatly aided when people agree on moral principles and relate their actions to those principles.

Chapter 6:   *MORAL DISPUTES*

Purpose of the Instructional Objective

At least some of the time we try to do what we think is right. Unfortunately, people differ in their conception of what is right and wrong: what ought to be and what ought not to be. When people differ it is sometimes possible to reach agreement through a process of arguing about morals. This process differs from the usual argumentation in that it consists of uncovering value assumptions and moral principles and questioning those principles. The first step in moral argumentation is to be able to spot when a moral dispute is present.

Instructional Objective

Students will be able to identify moral justifications versus motives, excuses, and rationalizations, and will be able to uncover value premises, identify corresponding moral principles, and evaluate those principles.

Sample Test Item

Directions

Indicate everything that is true in each example using these choices:

A)   An implicit value premise is underlined.
B)   An explicit value premise is underlined.
C)   A factual premise is underlined.
D)   A value judgment is underlined.

Example

If a movie shows violence in a favorable light it is not worthwhile. <u>The movie shows violence in a favorable light</u>, so it is not a worthwhile movie.

Answer

C

Organizer - Facts versus Value

Morality is concerned with not only what is, but also with what ought to be. In addition to using factual premises, moral reasoning uses value statements.

CONCEPT: Value Statements and Factual Statements

A value statement makes a claim about the worth of an object, experiences, ideas, actions, rules for living, etc. Typical value statements claim that something is good or bad, better or worse, ought to be or ought not to be.

A factual statement indicates what is or what exists. Factual statements include reports, descriptions, etc.

Note:

Not all value statements deal with morality. For example, "I have a good car" is not a moral claim. Generally, moral statements attribute a positive or negative value to human behavior. Morality relates to the control of behavior.

Different societies, and different philosophers have alternative views on how much behavior should be controlled. For example, in one society whether or not a person wears shorts in public might be considered a moral issue while in another society it is not a moral issue. On one extreme, there is the view that a person should be left alone so long as they do not physically hurt other people, and on the other extreme, there is the view that society should always be striving to help perfect people.

Exercise 6. 1   Directions: Indicate what is true of each example using these choices:

(A)   It is a value statement.
(B)   It is a factual statement.

1.   There is nothing more important than human life. ....___

2.   A person cannot live if his or her head is cut off. ....___

3.   That house is a good buy for the money. ....___

4.   Murder is wrong. ....___

5.   Heroin is illegal in the United States. ....___

6.   Privacy should be respected. ....___

7.   John Dunbar is a convicted murderer. ....___

8.   Every woman should have a free choice in the matter of abortion. ....___

9.   Abortion should be outlawed. ....___

10. Abortion is legal in the United States. ....___

11. Teenage pregnancy is increasing. ....___

12. This is good ice cream. ....___

13. It is wrong for teenagers to become alcoholics. ....___

14. Teenage alcoholism is increasing. ....___

15. We should have more education on alcoholism. ....___

16. If caught, prostitutes can be put into jail. ....___

17. The Bible is a bestseller of all time. ....___

18. The Bible should be a bestseller. ....___

19. My life is worth more than anything. ....___

20. Loyalty is the best thing. ....___

Most often people cite moral views to justify actions they have done or things that they want us to do. People also use other types of statements which can be confused with justifications. Rationalizations, motivations, and excuses are not the same as moral justifications.

CONCEPT: Justification, Rationalization, Motivation, and Excuses, and Explanations

Justify: To provide or show to be just, to vindicate, to pronounce free from guilt or blame. Justifications are reasons for accepting normative views. When arguments are presented to support our value judgments, arguments that use value premises, we are presenting justifications.

Rationalize: To attribute one's actions to rational and creditable motives without adequate analysis of the true motives. Rationalizations are not the real reason that person has done the action. Typically, a person using a rationalization chooses to act first and then seeks a reason for the action later.

Motivate: Motivations are basic forces that influence our actions such as hunger, fear, rage, and hope. If a person says "I did the action because I was afraid," they are telling us what motivated them to do the action. Mentioning a motivation does not give us a reason to conclude that the action taken was right, although it may provide an excuse for doing the action.

Excuses: Excuses are used when we do not want to be held responsible for an action. We present a reason why we do not feel that we are responsible. When people present excuses, they do not deny the truth of a moral claim -- they are claiming that they are exempt from doing what they should have done. An example would be a person saying, "I missed class because I was in the hospital with broken bones due to a serious automobile accident."

Explain: An explanation tells about an effect in terms of the causes. To explain is to make something clear to the mind.

Note: Some of these may go together. For example, if a person lost in a forest breaks into a cabin for food because they are starving, the motivation of hunger may provide an excuse for stealing.

Directions: Indicate what is true of each example using these choices:
   A)   It is a justification.   D)   It is an excuse.
   B)   It is a rationalization.   E)   It is an explanation.
   C)   It is a motivation.   F)   It is two of the above.

1.   Officer, I thought the speed limit was fifty-five, not forty-five, because that is
     the usual speed limit. I never would have gone over the limit had I known. ____

2.   I was speeding because I was excited about seeing my parents soon. ....____

3.   The speed limit really should be changed to fifty-five. It is quite safe to do
     fifty-five on this road. ....____

4.   I didn't go to practice because I didn't know there was one. ....____

5.   I get bored by driving and vary the speed to wake myself up. ....____

6.   I don't believe in gambling. One could lose all of his or her money. ....____

7.   I gambled because I was desperate to obtain some cash. ....____

8.   I ate the cake because I was hungry. ....____

9.   I didn't know that the meeting was at two p.m. It had been scheduled for
     three p.m. ....____

10.   I was so angry that I kicked the dog. ....____

11.   He says he did not have his homework because his dog ate it, but we
      know that he does not have a dog. ....____

12.   I was late to class because my watch broke. ....____

13.   I was so hungry that I cut class in order to get something to eat. ....____

14.   I did not get up on time this morning because I did not feel like it. ....____

15.   John does not eat fried food because fried food makes him ill. ....____

16.   I lied because I was afraid. ....____

17.   I didn't try to do well in the course because the material we were supposed
      to read probably wasn't interesting anyway. ....____

18. Water can be a fuel if hydrogen gas is separated from oxygen and then burned. ...._____

19. She could not slow down because she was worried about her children being home alone. ...._____

20. I didn't come to class because I don't like the teacher. ...._____

Exercise 6 - 3

Directions: Imagine the following situation -- Mr. Michael Anthony, the man who hands out the million dollar checks for an eccentric millionaire, has just given you a check. Looking over the check you notice that the middle name on it is not yours, but you want to keep the money anyway.

1. Offer an excuse for keeping the money.

_____

_____

_____

2. Rationalize your decision to keep the money.

_____

_____

_____

3. State your motivation for keeping the money.

_____

_____

_____

4. Provide a justification for your keeping the money.

_____

_____

_____

Exercise 6 - 4

Directions: Examine each passage below and determine the author's purpose. Use the concepts already defined—Justify, Rationalize, Motivate, Excuse, and Explain, and consider also the following concepts:

Describe: To represent by words, to give an account of, to trace or provide an outline of, as to describe a circle.

Persuade: To induce one to believe or do something.

1. You express a great deal of anxiety over our willingness to break laws. This is certainly a legitimate concern. Since we so diligently urge people to obey the Supreme Court's decision of 1954 outlawing segregation in the public schools, it is rather strange and paradoxical to find us consciously breaking laws. One may well ask, "How can you advocate breaking some laws and obeying others?" The answer is found in the fact that there are two types of laws: There are just and there are unjust laws. I would agree with St. Augustine that "an unjust law is no law at all." I hope you see the distinction I am trying to point out. In no sense do I advocate evading or defying the law as the rabid segregationist would do. This would lead to anarchy. One who breaks an unjust law must do it openly, lovingly, and with a willingness to accept the penalty. I submit that an individual who breaks a law that conscience tells him is unjust, and willingly accepts the penalty by staying in jail to arouse the conscience of the community over its injustice, is in reality expressing the very highest respect for law.

Reverend Dr. Martin Luther King
"Letter from Birmingham City Jail"

2. It may be suggested that one's reward should depend on how much one needs; the one who needs the most should receive the most. Some would doubt that need should be used as a criterion at all, and some would say that it should be used only sparingly or in extreme situations; but few would suggest that it should be the only criterion. If it were, work would soon come to a standstill, and there would be nothing left with which to reward anyone. Doubtless even those who are able to work but refuse to do so ... should not be allowed to starve and if possible should be sent to a psychiatrist ... But at least, reward cannot be based entirely upon need. If non workers were rewarded as much as workers, who would desire to work? Some doubtless would, but would then have to work much harder, to make up for the large mass of the indigent.

John Hospers, Human Conduct

3. We hold these truths to be self-evident, that all men are created equal; that they are endowed by their Creator with certain inalienable rights; that among these are life, liberty, and the pursuit of happiness. That, to secure these rights, governments are instituted among men, deriving their just powers from the consent of the governed; that, whenever any form of government becomes destructive of these

ends, it is the right of the people to alter or to abolish it, and to institute a new
government, laying its foundation on such principles, and organizing its powers in
such form, as to them shall seem most likely to effect their safety and happiness.
Prudence, indeed, will dictate that governments long established should not be
changed for light and transient causes; and, accordingly, all experience hath shown
that mankind is more disposed to suffer, while evils are sufferable, than to right
themselves by abolishing the forms to which they are accustomed. But, when a long
train of abuses and usurpations, pursuing invariably the same object, evinces a design
to reduce them under absolute despotism, it is their right, it is their duty, to throw off
such government, and to provide new guards for their future security. Such has been
the patient sufferance of these colonies, and such is now the necessity which
constrains them to alter their former systems of government. The history of the
present King of Great Britain is a history of repeated injuries and usurpations, all
having, in direct object, the establishment of an absolute tyranny over these states.

The Declaration of Independence

Justifications are susceptible to rational criticism. Arguments for moral claims may succeed or fail. Arguments may be valid or fallacious. This is apparent when you consider the general form of moral arguments.

CONCEPT: Value Judgments

Value judgments are value statements that occur in the conclusion of an argument. A value judgment is the conclusion of an argument that contains at least one value statement and at least one factual statement.

Example

An example of an argument that has as premises one value statement and one factual premise is:

Premise #1:   If a law degrades human personality, then it is an unjust law.

Premise #2: Segregation statutes degrade human personality.

Value Judgment: Segregation statutes are unjust laws.

Note: Value Conclusions Require Value Premises

It is generally accepted that whenever there is a value conclusion there must be a value premise. If this were not so, it would be possible to go from facts to values, from statements of what is to statements of what ought to be. You are welcome to try to derive a value statement from all factual statements. Many have tried to do this, but in each case a value assumption has been uncovered.

93

CONCEPT: Implicit vs. Explicit Value Premises

Value premises are not always stated. If they are stated, they are explicit. A value premise that is not stated but is assumed is called an implicit value statement. To discover implicit value premises just fill in the missing premises that are needed to make an argument complete.

Example

You promised that you would help me; so you should help me.

Procedure

Ask yourself,
1.   Is there a value judgment?
2.   If yes, then, is there an explicit value premise? If not, what is the implicit value premise?

Note that whether the person promised is a factual issue. That means that we have a factual premise and a value conclusion. A value premise is missing. The complete argument is:

If a person promises, then they should keep their promise.

You promised that you would help me.

_____

So, you should help me.

Exercise 6 - 5   Directions: Indicate what is true of each example using these choices:
(Note that whenever there is an implicit value premise the answer will be two of the above.)

         A)   An explicit value premise is underlined.
         B)   There is an implicit value premise.
         C)   A factual premise is underlined.
         D)   A value judgment is underlined.
         E)   Two of the above.

1.   Reading Shakespeare won't help a person make money, <u>so reading Shakespeare is a waste of time.</u> ....____

2.   If something is not fun then it is worthless. <u>Memorizing names is not fun</u>. Therefore, memorizing names is worthless. ....____

3.   <u>I do not find mathematics interesting.</u> Therefore I should not study mathematics. ....____

4.   <u>All people did not have the right to vote in South Africa</u> under apartheid. Any political system that does not afford rights to all those within that system is unjust. Therefore, the political system of South Africa was unjust. ....____

5.   <u>You said you would help me with my homework,</u> so you should help me. ....____

6.   <u>War involves killing</u> and if an activity involves killing then it is immoral. Therefore a person should not participate in the making of war. ....____

7.   <u>All citizens have a duty to protect their country from attack.</u> The way to protect the country from attack is to support having a strong military Therefore it is the duty of every citizen to support having a strong military. ....____

8.   What you did hurt other people, so <u>it was not a good thing to do</u>.............____

9.   Everyone should have the same amount of money. <u>Some people have more money than others.</u> Therefore the wealth should be redistributed.................____

10. <u>Having universal health insurance would make Americans more secure</u>, so it is the moral thing to do. ....____

Once you have uncovered value premises, you can argue that they should either be accepted or rejected. Value premises are part of value systems that individuals hold. Particular values must be consistent with general principles that we accept. After values have been stated explicitly, the next task is to find a relevant moral principle that is consistent with that value statement. Some of the most commonly accepted moral principles are presented here. That the principles are popular does not, of course, mean that they are correct. In fact, the first two principles presented cannot both be held consistently. The next section will discuss ways of judging between principles.

CONCEPT: The Principle of Equality or Equal Distribution

The Principle of Equality states that everyone should have the same amount of goods, privileges, etc.

CONCEPT: The Principle of Equity

The Principle of Equity states that everyone should have the same opportunity. The ultimate distribution may be unequal, but it is fair if everyone had an opportunity to get the larger share, the better grade, etc.

CONCEPT: The Principle of Relevant Difference

The Principle of Relevant Difference states that everyone should have same unless there are relevant differences between people. In a hiring situation, it is just to pay one person more than another only if there is a relevant difference between the people, such as that one has more education than the other. Sex and race differences are not usually thought to be relevant to the performance of a job. Thus, this principle rules out discrimination.

CONCEPT: The Utilitarian Principle

> The Utilitarian Principle states that the actions and distribution schemes should be judged according to whether or not they further the goal of having the greatest good for the greatest number of people.

CONCEPT: The Categorical Imperative

> The Categorical Imperative states that you should act so that the maxim of your action could become a universal law. This maxim rules out acts such as killing because a person could not consistently wish that everyone would engage in killing.

Exercise 6 - 6

Directions: Indicate what is true of each example using these choices:

      A)   The equality principle is being used.
      B)   The equity principle is being used.
      C)   The principle of relevant difference is used.
      D)   The Utilitarian principle is being used.
      E)   The Categorical Imperative is being used.

1.   You should not do an action that you would not want others to do. ....____

2.   If a teacher gives a higher grade to one student than to another, then there should be a reason for the difference in the grades. ....____

3.   Every student should have an opportunity to earn high grades. ....____

4.   All students should receive the same grade. ....____

5.   Poor students should be given state aid to go to college so that everyone has the same opportunities. ....____

6.   Everyone working in our co-operative enterprise should receive the same pay. ....____

7.   All students should be graded on the work they do. ....____

8.   The student with the higher score should receive the higher grade. ....____

9.   If two people do substantially the same amount, then they should be paid the same amount. ....____

10. It is best to have taxes reallocate wealth from the rich to the poor, because the poor benefit from getting extra money more than the rich do. ....____

Organizer - Identifying Moral Bases & Alternative Bases for Morality

When principles are questioned, it is necessary to examine the alternative sources that people claim for their moral views. Certain principles follow from certain conceptions of morality. The utilitarian principle, for example, is related to the consequentialist view of morality which holds that we should judge on the basis of consequences. Other views hold that judgments should not be based on consequences. Our choice of value system is ultimately based on the choice of how we base our ethical systems. Some of the alternative bases suggested for morality are presented here, along with potential problems with each view.

CONCEPT: Theological Basis

According the Theological Basis, God reveals moral rules. People who accept this view believe that they have a firm foundation for their views. A problem with the theological view is that it assumes that we know that God exists and what God wants us to do.

CONCEPT: Societal Approach

According to the Societal Approach, the proper moral rules are the ones accepted by most members of society. A problem with this view is that we would not want to accept all the practices that have in fact been accepted by the majorities. After all, genocide was accepted in Hitler's Germany and Stalin's Russia respectively, and slavery was once accepted in the United States.

CONCEPT: Consequentialist

A consequentialist believes that acts, rules and practices should be judged according to their usefulness in bringing about the satisfaction of desires. On this viewpoint, no act is intrinsically good or bad. One problem is accounting for obligations, such as to keep promises when consequences indicate that it would be better to break a promise.

CONCEPT: Intuitionist

According to an intuitionist, we have a special intellectual capacity to know what rules and actions are proper. Problems arise when accounting for differences in moral views.

CONCEPT: Naturalist

A naturalist holds the view that morals are part of the natural order. One problem of this view is that it would appear to confuse facts with values.

Exercise 6 - 7

Directions: Indicate what is true of each example using these choices:
- A) The Theological view is assumed.
- B) The Societal view is assumed.
- C) The Consequentialist view is assumed.
- D) The Intuitionist view is assumed.
- E) The Naturalist view is assumed.

1. Everyone is doing it, so it must be all right. ....____

2. We see in nature a pattern of the survival of the strongest. So it is right that the strong survive and the weak perish. ....____

3. The commandment says to honor thy father and mother, so it is right to do what your parents tell you to do. ....____

4. If one considers the cost to society and to the mother of having an unwanted baby, then it is apparent that abortion is sometimes the best choice. ....____

5. No one should steal because stealing threatens the security of our right to own property. ....____

6. God states that "Thou shalt not kill," so we are not supposed to kill. ....____

7. Stealing is condemned by most people, so it must be wrong. ....____

8. I feel that this thing you want me to do is wrong and I trust my feelings more than I trust you. ....____

9. People should be rewarded only according to what they have done in the past or are doing in the present. This is what satisfies our sense of justice. ....____

10. It is right to hire people of one ethnic group over people in another if it serves a social purpose in balancing professions, so that the proportion of people from ethnic groups in professions matches the proportion of people from ethnic groups in the general population. ....____

Chapter 7: DEDUCTIVE MOLECULAR ARGUMENTS

Purpose of the Instructional Objective

Once a person has determined that an argument is being offered and that the argument offered is a deductive argument, he or she needs to test the argument to see whether it is acceptable. There are different types of deductive arguments and each has to be tested by a different method. This chapter is concerned with the evaluation of deductive molecular arguments.

Instructional Objective

Students will be able to determine whether deductive molecular arguments are valid or invalid.

Sample Test Item

Directions

Indicate everything that is true in each example using these choices:

A) Valid
B) Invalid

Example

If John is in Boston then he is in the United States. John is the United States. Therefore, John is in Boston.

Answer

B) Invalid

Behavioral Objective

Given a mixture of examples of arguments some of which are molecular arguments and some of which are not molecular arguments the student will be able to identify the molecular arguments.

Given that parts of molecular statements are true or false, the student will be able to determine whether the entire molecular statements are true or false.

Given mixed examples of arguments, the student will be able to demonstrate whether each argument is valid or commits one of the following fallacies: affirms a disjunct, affirms the consequent, or denies the antecedent.

CONCEPT: Atomic and Molecular Statements

Molecular statements have parts and the truth of the whole is related to the truth of the parts. Note that the notions of part and whole are relative. In the example *"John is at the party and Mary is at the party"* the statement *"John is at the party"* is a part where the whole is *"John is at the party and Mary is at the party."* But in the statement *"It is not the case that John is at the party and Mary is at the party"* the statement *"John is at the party and Mary is at the party"* is only a part where the whole statement is *"It is not the case that John is at the party and Mary is at the party."*

Simply put, a molecular statement has parts which are statements and the truth of the whole is related in a truth functional way to the truth of the parts.

CONCEPT: Truth Functors

Molecular statements are statements with the terms "or," "and," "if - then," and "not." These terms are called truth functors because they indicate a truth functional relation between the part of a sentence and the entire statement.

John is at the party

It is not the case that John is at the party.

Note:

There is a relationship between the truth of the statement "John is at the party" and "It is not the case that John is at the party." The relationship is such that the whole statement has the opposite truth value of the part.

CONCEPT: Conjunctions, Disjunction, Conditional and Negation

Memorize the following vocabulary:

Conjunction "p and q" has conjuncts p, q
Disjunction "p or q" has disjuncts p, q
Conditional Statement:
    "if p then q" has an
    "if" part and a "then" part.

Memorize the following:

Conjunction	
∧ p and q conjunct	All conjuncts must be true for a conjunction to be true.
Disjunction	
∧ p or q disjunct	One or more of the disjuncts must be true for a disjunction to be true. The only time a disjunction is false is when all disjuncts are false.
Conditional	
If p then q antecedent consequent	All cases are true except when the "if" part is true and the "then" part if false. The "if" part is called the antecedent and the "then" part is called the consequent. The antecedent cannot be true with the consequent false, all other cases are true.
Negation	
not p	The negation of a statement has the opposite truth value of the original statement.

CONCEPT: Special Symbols

Special symbols are used in symbolic logic for the molecular connectors. These symbols are:

and	or	if ... then	not
&	v	⊃	~
p & q	p v q	p ⊃ q	~ p

Note:

It is important to separate talking about a disjunction and talking about a disjunct in a disjunction. Compare the following two expressions:

1.   not (A or B)   2.   not A or B

In the first expression, the disjunction A or B is being negated while in the second expression only the disjunct A is being denied. Supposing that A and B are both true statements, it would be the case that "not (A or B)" would be a false statement while "not A or B" would be a true statement.

Note:

The order in a conditional statement is important because the expression "if p then q" may be true when the expression "if q then p" is not true. When "if p then q" and "if q then p" are both true, the expression is called "biconditional." This is read "if and only if."

Note:

The disjunction "p or q" was defined as true when one or all the disjuncts are true. This definition includes the case of both disjuncts being true and is called the "inclusive or." Sometimes in English we use the word "or" in an exclusive sense as when we say "a person is either male or female," in which case we mean either one or the other and not both. Unless indicated otherwise, such as by stating "either ___ or ___," the word "or" is to be interpreted as inclusive.

Note:

Definitions of terms must be memorized. Don't try to reason them out. These definitions are conventions. There is no reason why "p or q" means that one or the other or both except that it has been agreed that "p or q" will be defined that way. They might have been defined differently, just as a "dog" might have been defined as something one sits on and a "chair" might have been defined as an animal that barks. It is more important that we all use terms the same way so that we use the same word for the same things. These are the conventions that have been agreed on for the use of terms and you must learn them so that we all use the words the same way.

Directions: Indicate what is true of each example using these choices:

A)   The statement is a conjunction.

B)   The statement is a disjunction.

C)   The statement is a conditional statement.

D)   The statement is biconditional.

E)   The statement is a negation.

1.   John lives here, but he receives his mail elsewhere. ....____

2.   John lives here or Mary lives here. ....____

3.   John lives here and Mary lives here. ....____

4.   John or Mary lives here. ....____

5.   John lives with Mary and Mary lives with Susan. ....____

6.   If John lives here, then Mary lives here. ....____

7.   John lives here if, and only if, Mary does. ....____

8.   Mary lives here provided that John does. ....____

9.   John lives here yet Mary lives here. ....____

10. John lives here only if Mary lives here. ....____

11. It is not the case that Sidney and John live together. ....____

12. Mary lives here unless John lives here. ....____

13. John does not live here.. ....____

14. "Mary lives here" is false. ....____

15. It is the case that Mary lives here if John does. ....____

16. Lovers are silly and donkeys are wild. ....____

17. Lovers are silly if and only if it is not the case that donkeys are wild. ....____

18. If you are wise, hard-working, and lucky, you will do well. ....____

19. If only she would look my way, then I could ask her out. ....____

20. I came, saw, and conquered. ....____

Exercise 7 - 2

Directions: Molecular statements allow people to make statements about the relationship between events, facts, etc. In order to explore relationships, molecular statements must be used. Now that the student is familiar with molecular statements it is desirable to apply the definitions to exploring some relationships. Write a brief essay on one of the following:

1.   What is the relationship of means to ends in Malcolm X's statement that Black people must be willing to use the ballot or the bullet? Was Malcolm X committed to violence?

2.   What should be the relationship between crime and punishment? Take C to mean that person A has committed a Crime and P to mean that Person A is punished by the law. Which of the following should hold?
     A) If C then P   B)   If not C then P   C)   If not C then not P

Exercise 7 - 3

Directions: When statements have the same logical function they should have the same grammatical form. This is also expressed by saying "use parallel expressions for parallel ideas." Correct the following:

1.   Susan likes reading, writing, and to swim in the lake.

     _____

2.   The team has always shown great respect and love of the game.

     _____

3.   Most students have the choice of vacationing or to go to school.

     _____

4.   The anthropologist spent a year in the desert, in the Alaskan wilderness, and the jungle looking for primitive tribes.

     _____

Directions: When ideas are not of parallel importance, it is necessary to subordinate one to the other. Rewrite these examples and subordinate ideas.

5.   Mary is talented, intelligent, and thinks she will be a lawyer.

     _____

6.   Burning the entire second floor, the fire started at midnight.

     _____

CONCEPT - Determining the Truth of Molecular Statements

Given that you know the truth or falsity of the parts in a molecular statement, you can figure out the truth or falsity of the molecular statement.

Sample Test Item

Directions

Given that A, B, and C are true and that X, Y, and Z are false, indicate whether the following statements are true or false.

Example

Not A and (B or Y)

Procedure

Not A and (B or Y)
    t   t   f

1. First, list what you know. You are given that A is a true statement, B is true, and Y is false.

       |
Not A and (B or Y)
    t   t   f

2. Next, notice the main molecular connector. The entire statement is a conjunction with "Not A" as one conjunct and (B or Y) as the other conjunct. Put an arrow above the main connector.

Not A and (B or Y)
not t | t   f
  | |
  f and   t

3. Evaluate the truth of each part of the larger molecular statement. "Not A" is the negation of something that is true, so it is a false statement. "(B or Y)" is a disjunction with one true disjunct and one false disjunct so it is a true statement because disjunctions are false only when both disjuncts are false, as shown to the left.

Not A and (B or Y)
not t | t   f
  | |
  f and   t
      |
     f

4. Next, work out for the entire statement which is a conjunction of the conjuncts "Not A" and (B or Y). Since a conjunction is true only when all conjuncts are true, the conjunction is false. The answer to the example is thus false.

Exercise 7 - 4

Directions: Given that A, B, and C stand for true statements and X, Y, and Z stand for false statements, indicate what is true of each example using these choices:

A) The statement is true.
B) The statement is false.

1. Not A or not Y .... ____

2. Not A and B .... ____

3. Not A and not B .... ____

4. A or not A .... ____

5. A and not A .... ____

6. (B and A) or not (B and not A) .... ____

7. (B and A) and not (B and not A) .... ____

8. Not B .... ____

9. A and not B .... ____

10. Not (A and B) .... ____

11. Not A or B .... ____

12. If A then X .... ____

13. If A then B .... ____

14. Not A or not X .... ____

15. If A then not B .... ____

16. (A and B) or (X and Y) .... ____

17. A or X .... ____

18. If (A and B) then (X or Y) .... ____

19. Not [(A and B) or (X or Y)] .... ____

20. A or B .... ____

Exercise 7 - 5   Name _____

Directions: Given that A, B, and C are true statements and X, Y, and Z are false statements,
              indicate what is true of each example using these choices:
        A) The statement is true.   B) The statement is false.

1.   (C v Z) & (Y v Z) ....____

2.   ~(B v X) & (Y & Z) ....____

3.   ~(Y ⊃ C) ....____

4.   ~(C v B) v ~(~X & Y) ....____

5.   ~[(A & B) v (X v Y)] & [(A & B) v (X v Y)] ....____

6.   ~[(~Y v Z) v (~Z v Y)] ....____

7.   ~A v X ....____

8.   [(B ⊃ Z) ⊃ B] ⊃ Z ....____

9.   X ⊃ (Z ⊃ A) ....____

10. ~B v C ....____

11. ~(B v X) & ~(Y v Z) ....____

12. ~[(C & B) v ~(~X & ~Y)] ....____

13. A⊃ (Y ⊃ C) ....____

14. [(X ⊃ Y) ⊃ B] ⊃ Z ....____

15. [(B ⊃ Z) ⊃ B] v Z ....____

16. A ⊃ (B ⊃ C) ....____

17. X ⊃ (B ⊃ C) ....____

18. ~A v [(B & (X v Y)] ....____

19. ~[(A & B) ⊃ ((A v X) ⊃ B)] ....____

20. (C v Z) & (Y v B) ....____

21. A v ~A ....____

22. (~B v C) ....____

23. (~B & Z) v (~A v Y) ....____

24. A ⊃ (B ⊃ C) ....____

25. X ⊃ (B ⊃ C) ....____

26. (X ⊃ Y) ⊃ Z ....____

27. A ⊃ (Y v C) ....____

28. (B & A) v (Y v C) ....____

29. (C v Z) & (X & A) ....____

30. (~X & B) v (~A & C) ....____

In general, if an argument of a particular form is valid, then any argument that has the same form is valid. Also if an argument of a particular form is invalid, then any argument that has that same form is invalid. Knowing this, all we have to do is the test several argument forms and record which are valid and which are invalid.

To provide a simple test I will use examples that are obvious. My examples are based on the relationship between clouds and rain and the fact that there must be clouds somewhere in the sky for it to rain.

First, test the forms that apply to conditional statements. The first premise will be a statement of the form "If rain then clouds." We will test four possible second premises to see which ones create valid arguments. Our possibilities are that the second premise might affirm the antecedent of the first, might deny the antecedent, or might either affirm or deny the consequent.

Affirming the Antecedent is a valid form:

If it rains then there are clouds
It rains
Therefore, there are clouds

This is a valid argument, so any argument of this form is valid. To describe the form we use names that show the relationship of the second premise to the first premise. This form is called "affirming the antecedent" because the second premise "It rains" affirms the antecedent of the first premise "if it rains then there are clouds." Any argument that affirms the antecedent is a valid argument, provided, of course, that its conclusion is not something totally unrelated to the premises.

Denying the Antecedent is an invalid form:

If it rains then there are clouds
It is not raining
Therefore, there are no clouds

Affirming the Consequent is an invalid form:

If it rains then there are clouds.
There are clouds
Therefore, it rains

Denying the Consequent is a valid form:

If it rains then there are clouds
There are no clouds
Therefore, it does not rain

Organizer: Additional Invalid Forms

We could go about mechanically testing all the different possible forms of arguments and record which are valid and which are invalid. It would, however, be adequate to just record which ones are invalid, since the arguments that are not invalid are valid. The easiest way to use the method of analogy is to just learn the few forms that are invalid. In conditional arguments there are two invalid forms: denying the antecedent and affirming the consequent. Now examine disjunctive arguments.

CONCEPT: Affirming a Disjunct

Affirming a disjunct is the only invalid form with the disjunction. To understand the example that follows remember that a disjunction is true when either one or both disjuncts are true.

The patient has pneumonia or asthma
The patient has pneumonia
Therefore, the patient does not have asthma

113

CONCEPT: Uncertain Relationship

If you cannot determine the relationship between the premises of a molecular argument, then the Fallacy of Uncertain Relationship is present. It is present in this example:

If it rains then John will get wet
It might rain
Therefore, John will get wet

The statement that "it might rain" does not quite affirm the antecedent. It is not certain what it does, so we say the fallacy of uncertain relationship is present.

CONCEPT: The Non Sequitur Fallacy

The fallacies considered so far -- denying the antecedent, affirming the consequent, affirming a disjunct, and uncertain relationship -- indicate problems in the premises. Another type of problem can appear in the conclusion. If there is no problem in the premises, but the conclusion brings up a topic that goes beyond the premises, then the fallacy of non sequitur is present. This is the same fallacy that was discussed in the chapter on informal fallacies. Here is an example of it in a molecular argument.

If the strikers win, then the company will lose money.
The strikers will win.
Therefore, the manager will resign.

Sample Test Item

Directions	Indicate what is true using these choices:
	A) Valid
	B) Denies the antecedent
	C) Affirms the consequent
	D) Affirms a disjunct
	E) Uncertain relationship
	F) Non sequitur

Example

If Mary is at the party then John is at the party.
John is at the party.

Therefore, Mary is at the party.

Procedure

1. Note the main connector. If the major premise is a conditional, then you know to check for affirms the consequent and denies the antecedent. (The example is a conditional.)

2. Check for fallacies that are relevant to the main connector. If it is not possible to tell the relationship between the premises, then the answer is uncertain relationship. (In the example the answer is that it affirms the consequent.)

3. After checking the premises, examine the conclusion to be sure that the conclusion is related to the premises. If it is not, then the fallacy of non sequitur is present.

Note:

The method of testing by analogy has some problems that symbolic logic overcomes. One problem is that an argument may appear valid because it affirms the antecedent yet be invalid because the conclusion is a non sequitur.

Exercise 7 - 6

Directions: Indicate what is true of each example using these choices:

- A) Valid
- B) Denies antecedent
- C) Affirms consequent
- D) Affirms disjunct
- E) Uncertain relationship
- F) Non sequitur

1. If Alice's car is at her house, then Alice is home. Her car is not at her house. Therefore Alice is not home. ....____

2. Susan will write if she is not angry. Susan has not written. Therefore, she is angry. ....____

3. If Jones belonged to a fraternity, Carol would be wearing a pin. Jones does not belong to a fraternity. Therefore, Carol is not wearing a pin. ....____

4. The tank is empty or the ignition is faulty. The tank is not empty, hence, the ignition is faulty. ....____

5. If Smith is not guilty, then he would welcome an inquiry. Smith welcomes an inquiry. It follows that Smith is not guilty. ....____

6. If Representative Brown runs for the State Senate, then Representative Mosley will not run for the Senate. Representative Brown is running for the State Senate. Therefore Mosley will drop out of politics. ....____

7. Paul is broke or he bought a new car. Paul is broke. It follows that he did not buy a new car. ....____

8. If Jefferson had made a touchdown, the fans would roar. The fans are not roaring. Therefore, Jefferson did not make a touchdown. ....____

9. If the Witmer Company is hiring more workers, then Mary will be hired. The company may be hiring soon. So Mary will be hired. ....____

10. Susan was present at the scene of the accident. Since if Susan were guilty then she would have been at the scene, it follows that Susan is guilty. ....____

CONCEPT: Necessary and Sufficient Conditions

A necessary condition is a condition which must be present if the effect is to occur. If N is a necessary condition for effect E, then E will not occur without N. Another way to state this is: if E occurs then N had to be present; or "If E then N" where E is an effect and N is a necessary condition for that effect.

A sufficient condition is a condition that will bring about the effect by itself. If S is a sufficient condition for effect E, then E will always occur when S is present. Another way to state this is: if S occurs the E has to occur; or "If S then E" where E is an effect and S is a sufficient condition for that effect.

In order to distinguish necessary and sufficient conditions you might first identify cause and effect and then see which position the causal factor fits into: "If *S* then E" or "If E then *N*."

Name _____

<u>Exercise 7 - 7</u>

Directions: Use the same choices as in the previous exercise.

1. A sufficient condition for Jane to be ill is that she have a fever. Jane has a fever. Hence, Jane is ill. ....____

2. For Bob to feel refreshed in the morning, it is sufficient that he get eight hours of sleep. Bob felt refreshed this morning. So he got eight hours of sleep. ....____

3. For Mary to play the piano well, it is necessary that she practice. Mary does not practice. Therefore, she doesn't play the piano well. ....____

4. If Joe studies hard, his studying is sufficient to insure that he does well. Joe did not do well, Therefore, Joe did not study hard. ....____

5. If Susan is a Freshman then it is necessary that she would have to be taking math. Susan is not taking math. Hence, she is not a Freshman. ....._____

Chapter 8:

STATEMENT VARIABLES AND ARGUMENT FORMS

Organizer

In order to evaluate the truth or falsehood of statements, to tell whether statements are equivalent or not, and to evaluate the validity of arguments, it is necessary to work on a higher level of abstraction. Instead of *particular* statements and arguments, statement variables and argument forms are discussed.

Instructional Objective

Students will be able to evaluate the truth or falsehood of statement variables, to tell whether statement variables are equivalent, and to determine the validity or invalidity of argument forms.

Sample Test Item

Directions

Indicate what is true of each example using these choices:

A) Valid   B) Invalid

Example

If Mary has a fever then she is ill. Mary is ill. Therefore, Mary has a fever.

Answer

B) Invalid

Behavioral Objective

After completing this chapter the student should be able to:

- Match arguments with argument forms

- Use truth tables to define molecular statements

- Indicate whether statements are tautologies, contradictions, or contingent

- Determine whether argument forms are valid using truth tables

- Spot invalid argument forms by finding the correct interpretation of truth values

Organizer - Argument Form

So far, capital letters in expressions such as A, B, X, etc., have been used to indicate particular statements which are all either true or false. When one uses particular statements, the results only apply to those statements. In addition to using capital letters which refer to particular statements, we can use small letters which refer to any statement whatsoever.

CONCEPT: Statements and Statement Variables

In the symbolic language being outlined, capital letters, such as A, B and C, refer to statements; while small letters, such as p, q, and r, refer to statement variables.

CONCEPT: Arguments

An argument is made up of statements. An example of an argument is:

If morality is a form of knowledge, then it can be taught.
Morality cannot be taught.

Therefore, morality is not a form of knowledge.

This would be symbolized as:

$$K \supset T$$
$$\sim T$$

$$\sim K$$

In this instance, the letter "K" stands for "Morality is a form of knowledge" and "T" stands for "Morality can be taught."

CONCEPT: Argument Forms

An argument form is made up of statement variables, such as the letters "p" and "q" which stand for any statement whatsoever. An example of an argument form is:

$$p \supset q$$
$$\sim q$$

$$\sim p$$

Organizer - Demonstration, Validity of Form

If we can prove that particular argument form is a valid form, then any argument that has the same form will also be valid. The form above is a valid argument form. Since "p" and "q" refer to any statements whatsoever, both of the arguments below have the same form as the argument form on the left.

Argument Form   Arguments that Match the Form

$$p \supset q$$ K $\supset$ T   (N & R) $\supset$ (S & Y)
$$\sim q$$ $\sim$T $\sim$(S & Y)
---------   ---------   ---------------------
-
$$\sim p$$ $\sim$K $\sim$(N & R)

If we can test argument forms and determine which forms are valid and which are invalid, then we can test arguments by matching them with argument forms.

CONCEPT: Matching Arguments with Argument Forms

A statement variable matches a statement provided that it results from substituting statement variables for statements consistently and preserving the same main statement connector. Examples:

R $\supset$ S   p $\supset$ q

Also: S & O   p & q

Sample Test Item

Directions   Indicate whether the second statement matches the first:

A) They match B) They do not match.

Example   (N & O) v L   (p & q) & r

Answer   B) They do not match because the main connector was changed.

Procedure   1. Examine the main molecular connector of the statements involved. Make sure that the main connector does not change.

121

2. Substitute like for like being sure to be consistent. This might involve substituting "p" for the first expression that appears, and "q" for the next expression, and then filling in with "p" and "q" consistently.

3. Other connectors besides the main connector must be preserved when a long expression is matched to a short one. You cannot arbitrarily change the connectors when you match. The expression "(A v B) ⊃ C" does not match "(p & q) ⊃ r" because the "or" connector cannot be arbitrarily changed to a conjunction.

Note:

It will sometimes be helpful in matching statements to statement variable to think of molecular statements, such as "(t v u)," as a unit.

Sometimes statement variables match statements letter for letter in which case it is an exact match, and sometimes a large expression matches a small expression in which case it is an inexact match. Either an exact or inexact match is still a match.

122

Exercise 8 - 1

Directions: Indicate what is true of each example using these choices.
A) The second statement is a match of the first.
B) The second statement does not match the first.

1. A v B .... p v q ....____

2. ~E ⊃ (F & G) .... p ⊃ q ....____

3. ~E ⊃ (F & G) .... ~p ⊃ (q & r) ....____

4. (K & L) v (M & N) .... (p ⊃ q) v (r & s) ....____

5. (R ⊃ S) v (T & ~U) .... (p ⊃q) & (r & ~u) ....____

6. [(X⊃Y) ⊃ X] ⊃ X............................ p ⊃ q ....____

7. C ⊃ (Y ⊃ C) .... p⊃ (q ⊃ r) ....____

8. ~B v X .... ~p & q ....____

9. [(B ⊃ Z) ⊃ B] ⊃ Z........................... p ⊃ q ....____

10. (Y v Z) & (Y v A) .... (p v q) & (p v q) ....____

Organizer - Truth of Statement Variables

How can you tell whether a statement form, which is made up of statement variables, is true or false? Before, we were able to tell whether statements were true or false by putting in the values that were given. For example, in "A v X," where A is a true statement and X is a false statement, we were able to find the truth value of the entire statement. But the "p v q" we don't know whether p is true or false, or whether q is true or false. Our method is to use a truth table to show all the possible interpretations for every statement variable.

CONCEPT: Truth Table

A truth table, truth matrix, or chart of possibilities is a device for showing all the possible truth values that propositions might have. A truth table shows the logical possibilities.

CONCEPT: Logical Possibilities

Logical possibilities are the possibilities that it is possible to express in a language, or that it is possible to conceptualize. Consider the proposition "Paul is at the party." This proposition might be true or false.

We will symbolize the proposition as "P" and write the truth values below it:

	P
1	t
2	f

These two lines are the logical possibilities. A line on the truth table represents a possibility or a possible state of affairs.

CONCEPT: Generation of the Chart of Possibilities

Consider the singular statement variable "p" and its truth table:

p
t
f

Now consider the singular statement variable "q" and its truth table:

q
t
f

Then consider the two together:

p
t
f

q
t
f

Where p and q are both in statements together as in "p and q" each one might be true or false while the other is true or false. While p is true, q might be either true or false, and while p is false, q might be either true or false.

A chart to show the values of p with the values of q would read as follows:

	p	q	
1.	t	t	p is true and q is be true
2.	t	f	p is true and q might be false
3.	f	t	p is false and q is true
4.	f	f	p is false and q is false or both are false

Note:

Since there are two truth values for each proposition, true or false, and two propositions there are four possibilities. Where there was only one proposition, P, there were two possibilities, true or false. Thus for one there were two possibilities, for two there were four possibilities.

125

CONCEPT: Use of the Truth Table to Define Terms

The truth table can be used to define the molecular terms. In discussing the meaning of terms we might refer to lines of the truth table. The symbol "v" is used for the "inclusive or." An "exclusive or" would have ruled out line 1 while the "inclusive or" rules out only line 4.

	p	q	p v q
1.	t	t	t
2.	t	f	t
3.	f	t	t
4.	f	f	f

- Definition of the term "v" (read "or") using the truth table p v q is true in all cases except where both p and q are false.

- Definition of the term "&" (read "and") using the truth table p & q is true only where both p is true and q is true. In all other cases it is false.

- Definition of the term "⊃" (read "horseshoe" or "if - then") p ⊃ q is true in all cases except where p is true and q is false.

Below are definitions of the molecular connectors using the truth table to define statements with each molecular connector.

Logical Possibilities

	p	q	p and q p & q	p or q p v q	if p then q p ⊃ q
1.	t	t	t	t	t
2.	t	f	f	t	f
3.	f	t	f	t	t
4.	f	f	f	f	t

CONCEPT: Truth Table Method for Statement Variables

We can determine the truth or falsity of statement variables such as "p & q" only by trying all the values for each statement variable. For the statement "p & q" we would have to first try to both p and q being true, then try for p being true with q false, for p being false with q true, and for both p and q being false. The possibilities for "p & q" are shown below:

p	q	p & q
t	t	t
t	f	f
f	t	f
f	f	f

Whether the entire statement "p & q" is true, or false, depends on what p is and what q is. We could say that the truth of "p & q" is contingent because the truth of the total statement depends on the truth of its parts "p" and "q."

To see how longer statement variable can also be shown on a truth table it is useful to compare the method used for statement variable with the method used for statements. The concepts are the same whether determining the truth of statement variables or determining the truth of statements. This will be illustrated by comparing a statement variable and a matching statement.

Statement	Statement Variable
~[(A ⊃ B) & (A v B)]	~[(p ⊃ q) & (p v q)]

In previous exercises, A, B and C were given as true statements and X, Y and Z as false statements. The steps that are taken to determine whether the entire statement is true or false are as follows:

127

Statement

$$\sim [(A \supset B) \ \& \ (A \lor B)]$$

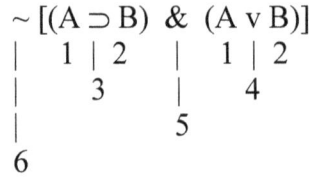

Procedure
1. Put in a truth value for the first letter.
   (A is true.)
2. Put in a truth value for the second letter.
   (B is true.)
3. Put in a truth value for A ⊃ B. (It is true.)
4. Put in a truth value for A v B. (It is true.)
5. Put in a truth value for (A⊃ B) & (A v B).
   (It is true.)
6. Put in a truth value for the negation of the above.
   (The answer is that the expression is false.)

Now examine how to evaluate the matching statement variable using the chart below. Because the truth or falsity of the variable is unknown, all possibilities must be tried by having four lines; one for each possibility. The columns on the chart correspond to the steps listed in evaluating the matching statement.

Statement Variable
$$\sim [(p \supset q) \ \& \ (p \lor q)]$$

Column 1 shows the value for the first letter, p.
Column 2 shows the value for the second letter, q.
Column 3 shows the value for p ⊃ q.
Column 4 shows the value for p v q.
Column 5 shows the value for (p ⊃ q) & (p v q).
Column 6 shows the value for ~[(p ⊃ q) & (p v q)].

	1.	2.	3.	4.	5.	6.
	p	q	p ⊃ q	p v q	(p ⊃ q) & (p v q)	~[(p ⊃ q) & (p v q)]
1.	t	t	t	t	t	f
2.	t	f	f	t	f	t
3.	f	t	t	t	t	f
4.	f	f	t	f	f	t

The final column, Column 6, shows the truth value for the entire statement. As can be seen on the chart, the entire statement is true on lines 2 and 4 -- when p is true and q is false and when p and q are both false. On lines 1 and 3 the statement is false.

Note:

Write notes to yourself. It is important when working out problems using a truth table to be clear about each step. It may be helpful to write notes to yourself such as the following notations in italics which explain the origin of each column. In the note above "p ⊃ q" it states that this is the same as the item in column 1 "horseshoe" the item in column 2.

Notes:		*1 ⊃ 2*	*1 v 2*	*3 & 4*	*~5*
1.	2.	3.	4.	5.	6.
p	q	p ⊃ q	p v q	(p ⊃ q) & (p v q)	~[(p ⊃ q) & (p v q)]

Note:

To set up for a column such as column 5, first find the main molecular connector. Place what is to the left of the main connector in a column to the left and what is to the right of it in a column to the right of that column.

CONCEPT: Tautology, Contradiction Contingent

The statement variable discussed in the last section, ~[(p ⊃ q) & (p v q)], is a contingent statement because the truth of the entire statement depends on the truth values of the variables in the statement. One some lines of the truth table the statement is true, while on other lines it is false. Whether the entire statement is true or false depends on which line of the truth table represents the actual facts. If in fact p is true and q is true then the statement is false. But, if in fact p and q are both false, then the statement is true.

Some statements are such that their truth or falsity is not dependent on the truth or falsity of their variables. Consider the statement p v ~p. In doing a truth table for this statement, only two lines are needed because there is only one variable, "p."

p	~p	p v ~p
t	f	t
f	t	t

129

Note that the truth of falsity of "p v ~p" does not depend on the truth values of "p." Its truth is due to its logical form. We don't have to know the value of p to determine its truth -- we know its truth from analysis. Statements which are true due to their logical form are called tautologies.

Now examine the statement "p & ~p" using a truth table:

p	~p	p & ~p
t	f	f
f	t	f

Note that the total statement is false regardless of what value "p" has. A statement which is false by virtue of its logical form is called a contradiction.

Sample Test Item

Directions   Indicate what is true for each item using these choices:
                          A) It is a tautology.
                          B) It is a contradiction.
                          C) Its truth is contingent.

Example   (p & ~p) ⊃ (q v ~q)

Procedure   1. Fill out a truth table:

Notes:		~1	~2	1 & 3	2 v 4	5 ⊃ 6
1.	2.	3.	4.	5.	6.	7.
p	q	~p	~q	p & ~p	q v ~q	(p & ~p) ⊃ (q v ~q)
t	t	f	f	f	t	t
t	f	f	t	f	t	t
f	t	t	f	f	t	t
f	f	t	t	f	t	t

2. Examine the truth or falsity of the entire statement.

130

3. If the statement is always true, then it is a tautology.

    If the statement is always false, then it is a contradiction.

    If there is a mixture of T's and F's in the column for the entire statement, then the statement is contingent.

Answer   It is a tautology.

Procedure for Setting Up Truth Tables

One way to think of how to set up problems using the truth table is to start at the right and work to the left. This is illustrated using the example above.

1. Note what the right hand column will be. In this example:

$$(p \mathbin{\&} \sim p) \supset (q \vee \sim q)$$

2. You know that if you had what is to the left of the main sign $\supset$ and what is to the right of that sign, then you can get the last column. Knowing this you know the next two columns to write, going from right to left.

next to next to last	next to last	last column
$(p \mathbin{\&} \sim p)$	$(q \vee \sim q)$	$(p \mathbin{\&} \sim p) \supset (q \vee \sim q)$

3. Continue in this same pattern. To get $(q \vee \sim q)$ you know that you need a "q" to the left and a "~q" to the right. Continue in this pattern remembering that you always have "p" as the first column and "q" as the second column, so you don't have to repeat these letters.

You will always have this	You need ~p for $(p \mathbin{\&} \sim p)$	You need ~q for $(q \vee \sim q)$	next to next to last	next to last	last column
p q	~p	~q	$(p \mathbin{\&} \sim p)$	$(q \vee \sim q)$	$(p \mathbin{\&} \sim p) \supset (q \vee \sim q)$
1 2	3	4	5	6	7 Columns

4. Write notes for yourself so you can check your work.

1	2	~1	~2	1 & 3	2 & 4	5 ⊃ 6

Notes

131

Directions: Indicate what is true of each example using these choices.
  A) The statement is a tautology.
  B) The statement is a contradiction.
  C) The statement is neither a tautology nor a contradiction.

1.   [p ⊃ (p ⊃ q)] ⊃ q ....____

2.   (p & q) & (p ⊃ ~q) ....____

3.   p ⊃ [p ⊃ (q & ~q)] ....____

4.   p & ~p ....____

5.   p ⊃ p ....____

6.   p ⊃ q ....____

7.   [p & (p ⊃ q)] ⊃ q ....____

8.   q v ~p ....____

9.   (p ⊃ q) ⊃ (q & ~q) ....____

10. (p ⊃ q) v (p ⊃ ~q) ....____

11. (p v q) ⊃ (p ⊃ ~q) ....____

12. p ⊃ ~~p ....____

13. [p v (p ⊃ q)] ⊃ q ....____

14. (p & q) & (p v ~q) ....____

15. (p & q) v ~(p & q) ....____

CONCEPT: Equivalent Statements

Two statements are truth functionally equivalent when they are both true or both false on all lines of the truth table. That the statements ~(p & q) and the statement ~p v ~q are equivalent can be demonstrated by doing a truth table that show these two expressions next to each other.

In making our chart of possibilities, it is important to be clear about what is being examined. In the chart below, only the last two columns are significant. The other columns may be thought of as auxiliary columns because they help us get to what we need to know. To end up with a column for ~(p & q), an *auxiliary* column for (p & q) must be established and to get that a column must be established for p and for q. The guides are written in to clarify the relationship between the columns.

Notes:		*~1*	*~2*	*1 & 2*	*3 v 4*	*~5*
1.	2.	3.	4.	5.	6.	7.
p	q	~p	~q	p & q	~p v ~q	~(p & q)
t	t	f	f	t	f	f
t	f	f	t	f	t	t
f	t	t	f	f	t	t
f	f	t	t	f	t	t

CONCEPT: Equivalences

Once you have tested statement variables and found them equivalent, you can test statements for equivalence by matching them. The sign to show that two expressions are equivalent is three lines. Below are some useful equivalences and their names:

$$\sim p \text{ v } \sim q \equiv \sim(p \text{ \& } q) \qquad \text{DeMorgan}$$
$$\sim p \text{ \& } \sim q \equiv \sim(p \text{ v } q) \qquad \text{DeMorgan}$$
$$p \supset q \equiv \sim q \supset \sim p \qquad \text{Transposition}$$

133

Exercise 8 - 3

Test the following to see if they are equivalent: *Yes*   *No*

1. p ⊃ q ≡ ~p ⊃ ~q   _____   _____

2. p ⊃ q ≡ ~q ⊃ ~p   _____   _____

3. p ⊃ q ≡ ~p v q   _____   _____

4. p & q ≡ q & p   _____   _____

5. p ⊃ q ≡ q ⊃ p   _____   _____

Exercise 8 - 4

1. What is the negation of this statement?
Williams is a mechanic and Johnson is a lawyer. ...._____

      A.   Williams is a mechanic and Johnson is not a lawyer.
      B.   Williams is not a mechanic or Johnson is not a lawyer.
      C.   Williams is not a mechanic and Johnson is not a lawyer.
      D.   None of the above.

2. What is the equivalent of this statement?
If Jones is not in Ohio, then he is in Millbrook. ...._____

      A.   If Jones is not in Millbrook, then he is not in Ohio.
      B.   If Jones is in Ohio, then he is not in Millbrook.
      C.   If Jones is not in Millbrook, then he is in Ohio.
      D.   If Jones is in Millbrook, then he is not in Ohio.

3. What is the negation of this statement?
In constructing an object, more connections mean more time, more
effort, and more strength. ...._____

      A.   In constructing an object, more connections mean more time,
            more effort, and less strength.
      B.   In constructing an object, more connections mean less time,
            less effort, and less strength.
      C.   In constructing an object, more connections mean less time,
            or less effort, or less strength.

Remember that in a valid deductive argument, if the premises are true then the conclusion must be true. The premises of a valid deductive argument provide conclusive reason for believing the conclusion. The concept of a valid deductive argument comes from the definition of a deductive argument. A valid deductive argument is one that is true to its definition. Therefore, an invalid deductive argument is one in which it is possible for the premises to be true and the conclusion false.

CONCEPT: Proving Invalid

In a valid deductive argument, if the premises are true then the conclusion must be true. If it is possible to find an interpretation of truth values for an argument form that has premises true and conclusions false, then that argument form is invalid. The following is an example of an invalid form:

if p then q	if F then T	T
q	T	T
--------------	----------------	---
p	F	F

If p is false and q is true then the premises are true and the conclusion is false, thus the argument is invalid.

Procedure for Showing Invalidity

Try listing the premises as true and the conclusion false. If you can succeed in finding an interpretation of truth value that makes the premises true and the conclusion false then the argument is invalid.

Note:

This method **cannot** be turned around to show that an argument is valid. Just because you can't spot an interpretation that makes it invalid does not prove it is valid. To prove an argument valid you have to exhaust every possibility to prove that is it impossible for the premises to be true and the conclusion false. Therefore a method is needed that shows **all** possibilities.

CONCEPT: Testing for Validity

In order to test for validity it is necessary to show all possibilities of truth values to demonstrate that there is no possibility of having premises true and conclusion false. If, after showing all possibilities (lines on a truth table), there is no line that has premises true and conclusion false, then the argument form is valid. If there is an interpretation (line) of truth values that has premises true and conclusion false, then the argument is invalid.

CONCEPT: Truth Table Method of Proving Validity/Invalidity

The truth table provides a method for testing argument forms such as the following:

1:    p v q
2:      p

~q

Note that premise 1 is "p v q," premise 2 is "p," and the conclusion is "~q." An argument form in this format has the conclusion located below the premises. However, in a truth table the conclusion is to the right of the premises. Picture it like this:

Premise 1	Premise 2	Conclusion
p v q	p	~q

In filling out the truth table to test the argument form, proceed one line at a time. In the first line try for p and q **both being true**:

possibilities		premises	conclusion	
p	q	p v q	p	~q
t	t	t	t	f

In the second case, try for p being true and q being false:

p	q	p v q	p	~q
t	f	t	t	t

Proceed in this way until the chart is filled in for all four possibilities:

136

p	q	p v q	p	~q
t	t	t	t	f
t	f	t	t	t
f	t	t	f	f
f	f	f	f	t

After all possibilities are filled in, examine the chart to see if all the premises are true and the conclusion is false on any line. If so, the argument form is invalid. If there are no such occurrences then you have proven that the argument form is valid.

Sample Test Item

Directions

Indicate what is true of each example using these choices:

    A) Valid   B) Invalid

Example

    p v q
      p

    ~q

Answer

B) Invalid

Procedure

A.   Method 1:

List the argument horizontally left to right --
1.   First, list all the statement variables: p, q, etc.
2.   List each premise.
3.   List the conclusion.

Construct a truth table as outlined above.

B.   Alternatively, a truth table may be filled in a column at a time instead of a line at a time. In filling in the chart, be careful to include enough columns and be clear about how to fill in each space. In the explanation provided below, columns of the chart are referred to by number.

The first line in the chart is notes that you write to yourself so that you can check your work.

possibilities   premise 1   premise 2   conclusion

Notes		*1 v 2*	*1*	*~2*
1.	2.	3.	4.	5.
p	q	p v q	p	~q

1. Column 1 and 2 are filled in mechanically to show the possibilities.
2. Column 3 is filled in from 1 and 2 because "p v q" is a disjunction with p a disjunct and q a disjunct.
3. Column 4 is filled in the same as column 1.
4. Column 5 is filled in reversing the truth values in column 2 because ~q is the opposite of q.

Once the chart is filled in, by either method, examine it to see if it is possible for the premises to be true and the conclusion false. Note which columns are the premises and which is the conclusion. In the second method above, column 3 represents the first premise, column 4 the second premise, and column 5 represents the conclusion. Ask yourself, "Is it possible for there to be a T in column 3 (premise one true) **and** a T in column 4 (premise 2 true) **and** an F in column 5 (conclusion false)?"

Note:

Remember that deduction works by listing possibilities and crossing off possibilities. If an argument is valid, the truth of its conclusion must follow. In a valid deductive argument, the possibility of the conclusion being false has been eliminated.

Exercise 8 - 5

Directions: Indicate what is true of each example using these choices:
    A)   The argument is valid.   B)   The argument is invalid.

1.   p ⊃ q
      p
    ----
      q   _____

2.   r ⊃ s
      s
    ----
      r   _____

3.   r v s
      s

    ~ r   _____

4.   o ⊃ n
     ~ o

    ~ n   _____

5.   t ⊃ s
     ~ s

    ~ t   _____

6.   q v s
     ~ q

    ~s   _____

7.   p ⊃ q
      q
    ----
      p   _____

8.   p v q
      p

    ~q   _____

9.   p & q
    ----
      p   _____

10.   s v r
      s

    ~r   _____

Chapter 9:

PROVING YOUR POINT WITH SYMBOLIC LOGIC

Purpose of the Instructional Objective

In order to evaluate complex arguments it is necessary to be able to symbolize arguments and to be able to use the tools of symbolic logic.

Instructional Objective

Students will be able to symbolize arguments from English to symbolic logic, to manipulate the symbols, and to evaluate arguments as valid or invalid by formulating proofs.

Behavioral Objective

After completing this chapter the student should be able to:

- Determine whether arguments are valid by using rules of inference

- Given a proof, fill in justifications

- Fill in missing steps of a proof

- Construct a proof

Organizer - Deduction Formalized

One can test arguments and record which forms are valid and which are invalid. It is then possible to compile a list of valid forms and invalid forms. For example, we know that the form "if p then q, p, therefore q" is valid. It is not necessary to go through all the steps the next time this form is encountered. Valid forms can be listed and used as "rules of inference."

CONCEPT: Names of Valid Argument Forms - The Rules of Inference

The valid argument forms are given names and called "rules of inference." The names used in this book are descriptive and tell whether a molecular connector is eliminated or introduced in the conclusion. The first two valid forms to be considered both have "and" as a connector.

p & q ------- p & E And Elimination	p q ------- p & q &I And Introduction

The first case begins with the connector "&" in the premise and eliminates it in the conclusion. In the other case, the connector "&" is added to the conclusion. The first form is known as "And Elimination" or "&E"; and the second form is called "And Introduction" or "&I."

The naming of the valid argument forms listed below follows the pattern of naming according to whether a connector is eliminated or introduced in the conclusion. This pattern of naming rules according to whether a molecular connector is eliminated or introduced is followed in the Partial List of Rules of Inference, except in the case where the traditional Latin name for the rule Modus Tollens is used to avoid having the same name for two rules.

PARTIAL LIST OF RULES OF INFERENCE

	Elimination	Introduction
&	p & q And Elimination ------- p &E	p And Introduction q ----- &I p & q
v	p v q Or Elimination ~p ------- vE q	p Or Introduction ---- p v q vI
⊃	p ⊃ q Conditional p Elimination ---------- q ⊃ E p ⊃ q Modus Tollens ~q ---------- MT ~p	

CONCEPT: Matching Arguments to Rules of Inference

Remember that if an argument matches a valid argument form, then that argument is valid. Any argument that matches a rule of inference is valid.

Sample Test Item

Example   Is the following argument valid?

$$(A \& B) \& (N \text{ v } O)$$
$$-----------------------$$
$$(A \& B)$$

Answer   Yes, it matches And Elimination, &E

CONCEPT: Logical Proofs

As long as valid argument forms are used an argument is valid. Long and complex arguments can be formed and tested by showing that each stage of an argument is composed of valid argument forms. If one can show that each step of an argument is composed of a valid argument form then one has shown that the total argument is valid. When every step in an argument is justified in this manner, it constitutes a logical proof.

Consider the following argument and the symbolization of it given below:

If Arthur bought two tickets to the theater, then he will not have enough money to last until payday. If Arthur is frugal, he will have enough money to last until payday. Arthur bought two tickets to the theater. Therefore Arthur is not frugal.

In symbolizing this argument, "T" represents "Arthur bought two tickets to the theater;" "M" means "he will have enough money until payday," and "F" means "Arthur is frugal."

$$
\begin{array}{lll}
1. & T \supset \sim M & \\
2. & F \supset M & \\
3. & T & / \therefore \quad \sim F
\end{array}
$$

Note that the first three lines represent the premises and that the conclusion is separated with a slash. To see whether the conclusion follows from the premises, show that the conclusion can be derived from the premises using only steps that match rules of inference. This proof takes two steps.

$$
\begin{array}{lll}
1. & T \supset \sim M & \\
2. & F \supset M & \\
3. & T & / \therefore \quad \sim F \\
4. & \sim M & \underline{1, 3 \supset E} \\
5. & \sim F & \underline{2, 4 \; MT}
\end{array}
$$

Read line 4 of the proof as, "~M" is derived from lines 1 and 3, and this matches the rule of inference called conditional elimination. To check whether this is correct, write line 4 as a conclusion and lines 1 and 3 as premises, and see if it does match conditional elimination:

$$
\begin{array}{ll}
1. & T \supset {\sim}M \qquad\qquad p \supset q \\
3. & \quad T \qquad\qquad\qquad\quad p \\
& \text{-----------} \qquad\qquad \text{--------} \\
& \therefore {\sim}M \qquad\qquad\quad \therefore q \qquad \supset E
\end{array}
$$

Line 5 of the proof is read as "F" is derived from lines 2 and 4, and this matches the rule called Modus Tollens (MT). Again, this is written out:

$$
\begin{array}{ll}
2. & F \supset M \qquad\qquad\quad p \supset q \\
4. & \quad {\sim}M \qquad\qquad\qquad {\sim}q \\
& \text{-----------} \qquad\qquad \text{----------} \\
& \therefore {\sim}F \qquad\qquad\quad \therefore {\sim}p \qquad MT
\end{array}
$$

This proves that the original argument is valid.

Sample Test Item

Directions

Fill in the justifications for each step in the following argument:

Example

Premise 1: A & E
Premise 2: (A v C) ⊃ D /∴ A & D (conclusion to .....
be reached)

3. A ...._____
4. A v C ...._____
5. D ...._____
6. A & D ...._____

Procedure

Remember the following:

1. Don't justify the premises -- they are given.
2. The slash indicates the conclusion being sought. This cannot be used in the proof but is stated and used as the last line of the proof.
3. Use any line above to prove any line below. Justification of a line proves it follows from the premises and, thus, it can be used. However, lines not justified cannot be used.

145

In seeking justifications, think of each step as a separate argument. In the sample test item above, work out the first justification as follows:

1. Think of line 3 as a conclusion.
2. Look above line 3 for statements that might serve as premises.
3. List an argument using line 1 as a premise and line 3 as a conclusion:

    A & E (line 1 as premise)

    $\therefore$ A   (line 3 as conclusion)

4. After setting up a step as an argument, examine the rules of inference to find a rule that would show that the argument is valid. In this case, And Elimination (&E) is the relevant rule.
5. Write on line 3 "by line 1, &E." This means, given line 1, line 3 is proved using the argument form And Elimination.

The complete proof for this problem is:

1: A & E
2: (A v C) $\supset$ D / $\therefore$ A & D
3. A ....   1, &E
4. A v C ....   3, vI
5. D ....   2, 4, $\supset$ E
6. A & D ....   3, 5, &I

Organizer

The materials we are covering build up step-by-step. Make sure you understand each step before moving on to the next. If you have any problem, review the step that comes before the one that bothers you. These are the steps:

1. Given a proof with justifications written in, understand why the justifications are correct. It is easier to read proofs than it is to formulate them.
2. Given a proof, write in justifications for each step.
3. Given a proof that is missing steps, write in the missing justifications.
4. Given an argument in English, write a proof and justifications.
5. Given a problem, write an argument, formulate a proof, and give justifications for each step of your proof.

Exercise 9 - 1

Directions: Fill in justifications for each step of these proofs.

I.
1. (M v N) ⊃ O
2. (O v P) ⊃ Q
3. P v M
4. ~P   / ∴ Q
5. M ...._____
6. M v N ...._____
7. O ...._____
8. O v P ...._____
9. Q ...._____

II.
1. K ⊃ L
2. L v M
3. (M & ~K) ⊃ (N & ~K)
4. ~L   / ∴ N
5. M ...._____
6. ~K ...._____
7. M & ~K ...._____
8. N & ~K ...._____
9. N ...._____

III.
1. C ⊃ ~D
2. ~C ⊃ (E ⊃ ~D)
3. (~F v ~E) ⊃ ~~D
4. ~F   / ∴ ~E
5. ~F v ~E ...._____
6. ~~D ...._____
7. ~C ...._____
8. E ⊃ ~D ...._____
9. ~E ...._____

IV.
1. (A v B) ⊃ (C & D)
2. A   / ∴ C
3. A v B ...._____
4. C & D ...._____
5. C ...._____

147

Exercise 9 - 2

Directions: Symbolize the following argument . Some lines are provided. Fill in the missing steps. In steps 7, 8 and 9 you must list the correct lines and rules. Use the names below the rules.

p ⊃ q _p_ q	p ⊃ q ~ q	p & q p	p q_ p & q	_p_ p ∨ q	p ∨ q ~p_ q
⊃E	M.T.	& E	&I	vI	vE

If Harmond is elected to the Retirement Fund Board then so will Johnson. Harmond will be elected to the Retirement Fund Board. Smith will also be elected to the Retirement Fund Board. In addition, if Johnson and Smith are both elected, then Bennford will be elected. If Bennford is elected, then he will be accused of embezzling retirement funds. Therefore, Bennford will be accused of embezzling retirement funds.

Use these symbols:
H = Harmond is elected to the Retirement Fund Board.
J = Johnson is elected to the Retirement Fund Board.
S = Smith is elected to the Retirement Fund Board.
B = Bennford is elected to the Retirement Fund Board.
A = Bennford is accused of embezzling retirement funds.

1.　　　　H ⊃ J
2.　　　　_____
3.　　　　S
4.　　　　_____
5.　　　　B ⊃ A　　　/ A
6.　　　　J　　　by I, 2　⊃E
7.　　　　_____
8.　　　　_____
9.　　　　A_____

Directions: Symbolize the following arguments and construct a formal proof for each.

1.   If Albert or Bob wins, then both Carl and David lose. Albert wins. Therefore, Carl loses.

2.   Whether people are moral is a function of their reasoning or of their passions. If of their reasoning, then morality can be taught. But morality cannot be taught. Therefore, morality must be a function of the passions.

3.   If nothing was stolen from the house, then robbery could not have been the motive of the crime. The crime was committed for robbery or it was vandalism. Nothing was stolen. Therefore the crime was vandalism.

Organizer - Equivalences Used in Proofs

> When two statements are equivalent, one can be substituted for the other in a proof as long as you cite the name of the equivalence.

CONCEPT: Equivalence: The General Formula

$$p \equiv q$$
$$p$$
$$\text{----------}$$
$$q \qquad \equiv E$$

> When two statements are equivalent and we are given that one is true, we may infer the truth of the other.

CONCEPT: Useful Equivalences

$p \supset q \equiv \sim q \supset \sim p$	Transposition	TRANS
$p \equiv \sim\sim p$	Double Negative	DN
$\sim(p \vee q) \equiv \sim p \, \& \sim q$	DeMorgan	DeM
$\sim(p \, \& \, q) \equiv \sim p \vee \sim q$	DeMorgan	DeM
$p \supset q \equiv \sim p \vee q$	Implication	Impl

CONCEPT: Commutation Laws

The following statements are equivalent. The biconditional statement may be read as "if and only if."

$$(p \mathbin{\&} q) \equiv (q \mathbin{\&} p)$$
$$(p \vee q) \equiv (q \vee p)$$

Name _____

Exercise 9 - 4

Directions: Fill in the justifications citing the relevant equivalences.

I. 1. (A v B) ⊃ ~C
    2. C   /∴ ~A
    3. ~(A v B)   _____
    4. ~A & ~B   _____
    5. ~A   _____

III. 1, (I v J) & (K v L)
      2. ~ K   /∴ L
      3.. _____
      4. _____
      5. L _____

II. 1. A ⊃ (B v C)
     2. ~B & ~C /∴ ~A
     3. ~(B v C)   _____
     4. ~A   _____

IV. 1. (N v O) & (R & S) /∴ R
     2. _____
     3. _____
     4. _____

Organizer - More Inference Rules

Two argument forms that are used in everyday life are hypothetical syllogism and constructive dilemma. These are presented below along with some suggestions on how to remember these rules of inference.

CONCEPT: Hypothetical Syllogism (H.S.)

$$p \supset q$$
$$q \supset r$$
$$\text{----------}$$
$$p \supset r \qquad \text{H.S.}$$

Hypothetical syllogism, also called chain argument, might be remembered by thinking of a chain of events. An event p causes an event q. Event q in turn causes event r. We then infer that p caused r.

CONCEPT: Constructive Dilemma

$$(p \supset q) \ \& \ (r \supset s)$$
$$p \ v \ r$$
$$\text{----------}$$
$$q \ v \ s \qquad \text{C.D.}$$

Constructive dilemma might be remembered by thinking of the "horns of a dilemma." On one horn is "p ⊃ q" and on the other is "r ⊃ s." Given that we will have to choose p or r, we will end up with q v s.

Exercise 9 - 5   Name_____

Directions: In each of the following examples, supply both the missing line and the justifications.

I. 1. R ⊃ ~S
    2. T ⊃ S / ∴ R ⊃ ~T
    3. ~S ⊃ ~T _____
    4. R ⊃ ~T _____

II. 1. (A ⊃ R) & (B ⊃ F)
    2. A v B
    3. (R ⊃ B) & (F ⊃ W) / ∴ B v W
    4. _____
    5. B v W _____

Organizer - Rules of Inference that add a premise

Two rules of inference involve adding a premise arbitrarily to see what adding that premise will produce. These rules are, Conditional Introduction (or "conditional proof") and Negation Introduction (or "*argumentum ad absurdum*.").

Conditional Proof	Argumentum Ad Absurdum

p   Conditional
        Introduction
q   ⊃ I
----- 	
p ⊃ q
	p Negation
	          Introduction
	q & ~ q ~ I
	--------- 	
	~p

CONCEPT: Conditional Proof

If adding the premise "p" makes it possible to derive "q", then "p ⊃ q" must be true. To see why this is so, imagine that "A ⊃ B" is given as true, then if A is true, we can derive B from it. We could not derive B from A unless we had A ⊃ B. In our Conditional Introduction rule of inference we start with the antecedent "p" and derive the consequent "q" and then conclude that "p ⊃ q" has to be true. We are entitled to do this because we could not derive q from p unless p ⊃ q were true. This rule is also called conditional proof.

CONCEPT: Proof by Contradiction

The rule called Negation Introduction in this text, which is also called *argumentum ad absurdum*, is based on the fact that we cannot derive a false conclusion from true premises in a valid argument. By using rules of inference in our argument, we insure that the argument is valid. If we assume a premise "p" and derive the false statement "q & ~q," then we know that the original statement "p" had to be false. The *argumentum ad absurdum* pattern of inference is used when a person assumes a view that he or she believes is false for a brief period in order to show the absurdity of that view.

Exercise 9 - 6

Directions: Indicate the type of argument: proof by contradiction, dilemma, etc.

1. On Friday, Mike had to attend school for four hours and do homework for six hours. Mike played softball for three hours, went to a party for four hours, and slept 8 hours. Therefore, Mike either did not attend class or did not study as needed because if he did, then there would be more than 24 hours in a day.

2. If I study for the test then I will have a terrible time. If I don't study then I'll get a low grade. I will either study for the test or I will not study for the test. Therefore, I will have a terrible time or I will get a low grade.

3. Suppose it is true that women are paid less than men to do the exact same work. Some claim that women are paid fifty nine cents to every dollar men get. Clearly it would be in the interest of every employer to hire women instead of men. A company that hired men could not compete with one that hired women and would soon go out of business. But men are able to find jobs. Therefore, women do not do the same work as men for less pay.

4. Euthyphro claimed to Socrates that he knew what piety meant and then proceeded to claim that an act is pious if and only if it is loved by the Gods and that an act is impious if and only if it is not loved by the Gods. Euthyphro then stated that what one God loves another God does not love. From these three premises Socrates proved that according to Euthyphro's definition of piety the same thing might be both pious and impious. (You can write out the proof of this.)

Chapter 10:

HANDLING "ALL" AND "SOME" IN SYMBOLIC LOGIC

Purpose of the Instructional Objective

In order to be able to handle arguments that use the words "some" and "all," we need to add quantifiers to symbolic logic.

Instructional Objective

Students will be able to symbolize statements and arguments from English to symbolic logic when the quantifiers "some" and "all" are present.

Behavioral Objective

After completing this chapter the student should be able to:

- Match English statements that contain the words "some" and "all" to their symbolic equivalents.

- Symbolize English sentences that indicate quantification.

- Fill in justifications in arguments that have quantifers.

- Formulate proofs for arguments that contain quantifiers.

Organizer - Quantification

The quantifiers "all" and "some" are used to refer to all the members of a category or to some of the members of a category. Symbols may be used instead of the words "all" and "some" to build our symbolic language so that it can handle quantifiers.

CONCEPT: Symbols for the Existential Quantifier -- Some

- The symbols x, y, z, etc., are variables.
- The symbol "$(\exists x)$" means "there exists at least one x"
- The expression $(\exists x)(Sx \& Qx)$ might be interpreted as "there exists an x such that it is a symbol and it is a quantifier"
- The expression $(\exists x) (Kxy)$ might be interpreted as "there exists an x such that x knows y"

CONCEPT: Symbols for the Universal Quantifier -- All

- The symbols x, y, z, etc., are variables.
- The symbol (x) means "for all x's" or "each thing x is such that"
- The expression (x)(Qx v ~Qx) might be interpreted as "for all x's, x is either a quantifier or x is not a quantifier"
- The expression (x)(Kxy) might be interpreted as "for all x's, x knows y" or as "every x knows y"

CONCEPT: Free and Bound Variables

A quantifier can be attached to either an entire statement or to part of a statement. When it applies to only part of a statement, parenthesis are included to indicate the scope of the quantifier. There can be different quantifiers in one expression, each with its own scope as in the example below. "At least one person is kind or all people would be unkind" might be symbolized as:

$$(\exists x)\ Kx)\ v\ (y)(\sim Ky)$$

In this example, x and y in Kx and ~Ky are bound because they refer to x and y in the parenthesis. This would be an example of variables that are bound.

Variables are not bound if they do not refer back to anything that is in a parenthesis. In the statement below x in Kx is not bound. The statement reads, "x is kind or all y's are unkind":

$$Kx\ v\ (y)\ (\sim Ky)$$

In the above example the x in Kx is called a free variable because it does not refer back to anything. It is still a variable, and we have no way of knowing whether an expression with a free variable in it is true or false.

Exercise 10 - 1

Name _____

Directions: Match these statements using the following choices only once each.

A) (x) (Sx ⊃ ~Px)   C) (∃x) (Sx & ~Px)
B) (∃x) (Sx & Px)   D) (x) (Sx ⊃ Px)

1. All S's are P's ....____

2. No S's are P's ....____

3. Some S's are P's ....____

4. Some S's are not P's ....____

Exercise 10 - 2

Directions: Match the English expressions with the symbolic statements, using each choice only once.

A) (x) (Mx ⊃ ~Fx)
B) Mx & Fy
C) (x) (Mx v Fx) & (x) ~(Mx & Fx)
D) (∃x) (∃y) (Mx & Fy)
E) (x) (Fx) v My

1. Every person is either male or female. ....____

2. Every person is female or y is a male. ....____

3. x is male and y is female. ....____

4. For all x's, if x is a male then x is not a female. ....____

5. There exists as least one person that is male and at least one person
   that is female. ....____

Exercise 10 - 3

Directions: Put each of the following into symbolic notation.

1. None but the strong survive.
    (Sx = x is strong; Vx = x survives)   _____

2. He who laughs last, laughs best.
    (Lx = x laughs last; Bx = x laughs best)   _____

3. There are no politicians who are not ambitious.
    (Px = x is a politician; Ax = x is ambitious)   _____

4. Some corporations do not act responsibly.
    (Cx = x is a corporation; Rx = x acts responsibly.)   _____

5. The only birds that are flightless are ostriches and emus.
    (Bx = x is flightless; Ox = x is an ostrich;
    Ex = x is an Emu; Fx = x can fly)   _____

6. The only animals I have in my house are cats.
    (Ax = x is an animal in my house; Cx = x is a cat)   _____

7. Horses exist but unicorns do not.
    (Hx = x is a horse; Ux = x is a unicorn)   _____

8. Only a fanatic would advocate violence.
    (Fx = x is a fanatic; Vx = x advocates violence.   _____

9. Everything that is enjoyable is either immoral,
    illegal, or fattening.
    (Ex = x is enjoyable; Mx = x is moral; Lx = x is legal;
    Fx = x is fattening)   _____

10. None think the famous unhappy but the famous.
    (Ux = thinks the famous unhappy; Fx = x is famous)   _____

CONCEPT: Quantitative Equivalencies

If two statements are equivalent then we can substitute one for another. The following equivalencies can be used in proofs.

Quantitative Equivalence

$\sim(x) Fx \equiv (\exists x) \sim Fx$

$\sim(\exists x) Fx \equiv (x) \sim Fx \qquad$ EQUIV

One way to remember these equivalences is to note that the negations seem to go through the quantifiers and that each time they do the quantifiers change (from either "all" to "some" or from "some" to "all").

CONCEPT: Quantification Rules

The following rules apply to the elimination and introduction of the universal and existential quantifiers.

Universal Elimination (x) E

$(x) Px$

Pa

From the universal statement (x) Mx, we can infer the instance Ma. An instance, or instantiation, is anything we get by eliminating the quantifier and inserting a name in place of a variable. From (x) Mx we can infer Mb as well as Ma. There may be many different instantiations. If all humans are mortal then Bob, Susan, George, and Mary are all mortal. Our first instantiation rule is a rule of inference called Universal Elimination or (x)E. This rule states that if something is true of all members of a category x, then it is true of each particular member of the category.

```
┌─────────────────────────────────────┐
│                                       │
│      Universal Introduction (x) I     │
│                                       │
│                  Pa                   │
│               --------                │
│               (x) Pa                  │
│                                       │
│        Provided that a refers to any  │
│      arbitrarily selected individual. │
│                                       │
└─────────────────────────────────────┘
```

While we can infer from "all" to "one" we cannot go from one to all, from "John Paul Getty is rich" to "all people are rich." The argument from one to all is valid only if we pick a feature that is true of all so that the one selected might be chosen arbitrarily. For example, from the statement "any person you select is mortal" we can infer "all people are mortal." This provides us with a rule of inference called Universal Introduction or (x) I.

```
┌─────────────────────────────────────┐
│                                       │
│      Existential Elimination (x) E    │
│                                       │
│               (∃x) Px                 │
│               --------                │
│                 Pa*                   │
│                                       │
│       *Provided that a is a variable  │
│        that has not been used before. │
│                                       │
└─────────────────────────────────────┘
```

The Existential Elimination, (∃x) E, rule is valid only with certain restrictions. Without the restrictions, we could infer from the statement "there exists a person who is poor" that Kennedy, Rockefeller, or John Paul Getty is poor. We can remove the quantifier (what we call instantiation) only if we replace the variable x with another variable that has not been used before. With this restriction, all we can infer from "there exists an X, such that X is poor" is that "someone is poor." We cannot give a proper name to that someone and that someone cannot be a person who was mentioned before.

From "John Paul Getty is rich" we can infer that "there exists at least one x such that x is rich."

CONCEPT: Proofs Using Quantifiers

The quantification rules and quantificational inferences can be added to our previous list of rules and equivalences to aid us in doing proofs. The rules and equivalences are repeated here so that you can follow as sample proof that uses them.

Quantificational Rules and Equivalences	
Universal Elimination (x) E (x) Px* ----------- Pa *cannot refer to a fictional being	Existential Elimination (x) E (∃x) Px ----------- Pa* *provided a is a variable not used before
Universal Addition (x) A Pa* ------------ (x) Px *provided a refers to any arbitrarily selected individual	Existential Addition (x) A Pa --------- (∃x) Px
Quantificational Equivalences - EQUIV ~(x) Fx ≡ (∃x) ~Fx ~(∃x) Fx ≡ (x) ~Fx	

Sample Proof

Directions   Fill in the justifications for each step.

Example   ^(x) Gx & (y) Gy   / ∴ Ga & ~Ga

Procedure   The following is correctly done:

1.   ~ (x) Gx & (y) Gy   / ∴ Ga & ~Ga
2.   ~ (x) Gx <u>1, &E</u>
3.   (∃x) ~ Gx   <u>2, EQUIV</u>
4.   ~ Ga   <u>3, (x)E</u>
5.   (y) Gy& ~(x) Gx 1. COMM
6.   (y) Gy   <u>1, &E</u>
7.   Ga   <u>5, (x)E</u>
8.   Ga & ~ Ga   <u>4, 7, &I</u>

162

Exercise 10 - 4

Directions: Fill in the justifications.

I.   1.   (x) (Nx ⊃ ~Rx)
      2.   (x) (Gx ⊃ Nx)   / ∴ (x) (Gx ⊃ ~Rx)
      3.   Ny ⊃ ~Ry ...._____
      4.   Gy ⊃ Ny ...._____
      5.   Gy ⊃ ~Ry ...._____
      6.   (x) (Gx ⊃ ~Rx) ...._____

II.   1.   (x) (Wx ⊃ Dx)
      2.   (∃x)(Wx & Ux)   / ∴ (∃x) (Ux & Dx)
      3.   Wa & Ua ...._____
      4.   Wa ⊃ Da ...._____
      5.   Wa ...._____
      6.   Da ...._____
      7.   Ua & Wa ...._____
      8.   Ua ...._____
      9.   Ua & Da ...._____
     10. (∃x) (Ux & Dx) ...._____

Exercise 10 - 5   Directions: Formulate proofs for the following valid arguments.

1.   All mathematicians are rational. Enrico Fermi is a mathematician. Therefore Enrico Fermi is rational.

2.   All wars are unjust, but some wars are understandable. Therefore, there are some unjust things that are understandable.

3.   All these art works are valuable. Whatever gets stolen must be paid for. Therefore, if everything valuable gets stolen, then all these art works will have to be paid for.

4.   There exists a thing that is disliked by any person. Every person dislikes something or other. Therefore, there is something that is disliked by all people.

5.   Every difficult problem is a challenge. Anyone who solves anything challenging deserves credit. Some clever person solved certain difficult problems. Therefore, some clever person deserves credit.

Appendix

Student Papers

These papers are printed here with the written permission of the papers' authors who agreed to have papers included without their names. The topics of the papers were chosen by the students and reflect their individualities. The collection begins with a paper entitled "Drug Abuse Should Be Stopped" which is followed by a critique of that paper. I think that both the paper and the critique are good examples of what might be done in assignments that involve writing and criticizing short argumentative papers. The other papers were included for many different reasons. Some are included because they were insightful, others were chosen because they provide examples of fallacies. Still others were chosen because they discuss controversial issues that might stimulate a response from other students. A complete list of titles appears below.

Contents by Title

A-1.   Drug Abuse Should Be Stopped

In recently published statistics, the National Commission on Marijuana and Dangerous Drug Abuse states that approximately 1.5 million Americans of high school age and 700,000 adults admit they've tried heroin at least once. This shows that teenagers are willing to experiment with heroin in increasingly large numbers. This means that youthful addiction to heroin will become a large problem.

At one point, society believed that education on the dangers posed by drugs was the best approach to discouraging their use. Unfortunately, this approach is not working. Some of the most well-informed young people today are doing drugs with total disregard for their future. Millions of young people are being lost to drug abuse. Many are highly talented people who thought "I can handle it," or "I can stop whenever I choose." Then they discover they can't. Nobody knows how many addicts there are in this country, but it is clear if not stopped we will be a drug-taking society.

A-2. Critique to "Drug Abuse Should Be Stopped"

In the author's paper she said that statistics by the National Commission on Marijuana and Dangerous Drug Abuse stated that 1.5 million students and 700,000 adults admit that they have tried drugs at least once. She claims that this shows that teenagers are willing to experiment with drugs, and that youthful addiction will become a large problem. She also said that many young people are being lost to drug abuse, and that nobody knows how many addicts there are in this country.

The paper had a good structure, but her argument was invalid. She missed the fact that the statistics showed the 1.5 million students and 700,000 adults admit only that they have tried drugs. Because the people tried the drug doesn't make them addicts. Many people who tried the drug and didn't like it may have stopped there.

The author also contradicted herself. She said that statistics show that 1.5 million students and 700,000 adults admitted they tried drugs, but farther down in her paper she said nobody knows how many addicts there are in this country. If this is true, then the statistics wouldn't be sufficient evidence that youthful addiction is becoming a large problem.

A-3. I Think Pot (Marijuana) Should Be Made Legal

To make a statement like this, one must first be experienced in the use of pot. When I say experienced, I don't mean getting high on it only once, I mean getting high every day or very often. I think pot should be legalized simply because I've been smoking pot for about five years and I do smoke often, and to my knowledge it hasn't affected me mentally, emotionally, or physically.

A-4. College Students Should Be Treated Individually

I feel that college students should be tested on individual and not group programs. In the course of a semester, a student may learn several different things about a subject, but may be interested in one thing stronger than he is in another. The more a student cares about something, the more attention he places on that particular thing, and the more he learns about it. A student should be tested individually on what he or she is most interested in.

Some instructors may argue that there are certain things that must be learned during a course before you are able to move on to the next course. But I feel that if the student has taken any special interest in a part of the course, he should be tested and graded on how well he handles this specific area. Students take more interest in one area than they do another, so they should be tested individually according to their subject interest.

When an instructor gives a test, he is asking for material that he feels everyone should have learned. I think this is bad. No two students may have the same interest, and therefore may not learn about the same things equally. So how can an instructor test students equally? I feel that individual testing is what we need and I also feel that individual testing would make the students more interested in learning. And if students worked harder in learning they would also make better grades.

A-5. Homosexuals Should Be Treated Like Anyone Else

I believe homosexuals should be treated like anyone else. If we treat others as equals, then we should treat homosexuals as equals. My three basic premises are these: first, as long as someone does not hurt others, they are moral; second, homosexuality is no worse than any other self-inflicted injury (like smoking); and third, any sexual act between consenting adults is acceptable and moral.

I argue that the love-making in homosexual relations is a sexual act between consenting adults and moral. I assume that if it is moral, it must not hurt others. Lastly, though homosexuality may not be fulfilling to a person psychologically, at worst it harms only the person himself, and is at worst self-inflicted harm, and thus no worse than other self-inflicted actions such as smoking.

A-6. Children Should Be Bussed to Obtain Racial Balance in School

I think children should be bussed in order to obtain racial balance. First, bussing is just about the only way that balance will be achieved. There have been a number of ways tried, but so far bussing is the only what that has done the job it was supposed to do, although many parents much more than the children dislike it.

One of the solutions to achieve racial balance that has been tried is freedom of choice, which ended up as follows: all black and no white; all white and no black; or a majority of whites and minority of blacks. Another solution tried was going to the school nearest you. This solution turned out as all black, all white, majority to black and a minority of whites, (which seldom occurred), or a majority of white and a minority of blacks. The latter was the case most often because white parents are unwilling to let their children go to an all black school.

Most of the parents have said that their children are in a sense being "used" to obtain racial balance in the schools, but isn't that what it is all about? Either way the children are going to be the victims. Some people say that they believe in integration but they do not think bussing is the right way to go about it. What they are failing to realize is that bussing is the only solution so far that has worked and is working. Therefore, if you believe in integration, and bussing is the only way it has been and possibly can be obtained, then way fight it?

A-7. Why Integration Won't Work

It is a belief of mine that true integration is a virtual impossibility. It is a fact that man is able to distinguish between the races. The fact that he is able to discriminate, causes him to discriminate against. I would argue that if man was unaware of any difference between the races, there would be no segregation. Since man has a nature that is geared toward classifying, categorizing, and discriminating, true integration can be a possibility in theory only.

167

A-8. Social Discrimination

I was talking to a friend the other day and she asked me if I regretted having not pledged a fraternity. Of course my reply was no, because I cannot see any good in pledging. She then said, "Well, suppose you were looking for a job and you ran across a fellow fraternity brother. The two of you get to talking and you discover that you and he are members of the same fraternity. On that basis, and that basis along, you get the job."

My argument is as follows: why should a fellow human being be a member of the same fraternity or of the same race, or of the same anything for that matter, to lend a helping hand to a fellow human being in need or deserving of a chance to prove himself worthy? The act of pledging in itself perpetuates this same concept of giving assistance on the basis of who they are; not what they are or abilities possessed.

Does mankind really need some group as small as a fraternity or some other social organization to relate to another different but equal human being? Will man ever see himself as one, as all being human and the same? Or will he be doomed to walk throughout eternity saying, "I'm American, I am better than you." or "I'm white and I am better than you." We label people as rednecks and bigots, capitalists and communists, Protestants and Catholics. Suppose these words were not in our vocabulary or, if we were blind to the fact that these differences exist, then we would be living ideally. However, these words do exist and they are a reality with which we must deal. People, most people, tend to see what they wish to see in others. They are stopped cold by nationality, religion, persona prejudices and cannot allow themselves to relate to others because of them.

My point is that fraternities do indeed carry on and uphold these prejudices. If you do not "belong", you are out, brother, out. Why is it so extremely difficult for people to see other people for what they are inside, to like and appreciate them for their abilities, talents and personal characteristics? They very act of pledging to a fraternity inhibits and rigidly forces a member to conform to a certain pre-cast mold which is difficult to break out of. I say live to your fullest individual potential and maintain an open mind.

A-9. There Is A God

I was talking to a friend of mine yesterday and he said there is no God. He believes if there was a God, He would have stopped war years ago. I strongly disagree with him. I don't think He is responsible for the predicament the world is in. If He corrected every mistake man makes, we would never learn for ourselves. He has to let things take their course for us to benefit from our mistakes.

I think there is a God because I was brought up to believe in Him from early childhood. Someone has to have control over things. According to my beliefs there are a number of things that make it impossible for there not to be a God. If there is no God, then who created man? Who created nature? Who creates day and night? The premises I have stated based on my beliefs and knowledge are true. Therefore my conclusion, "There is a God," has to be true.

168

www.ingramcontent.com/pod-product-compliance
Lightning Source LLC
Chambersburg PA
CBHW081152270326
41930CB00014B/3129